Engineering Statistics

Edward B. Magrab

Engineering Statistics

An Introduction

Edward B. Magrab
University of Maryland
College Park, MD, USA

ISBN 978-3-031-05012-1 ISBN 978-3-031-05010-7 (eBook)
https://doi.org/10.1007/978-3-031-05010-7

This Springer imprint is published by the registered company Springer Nature Switzerland AG
The registered company address is: Gewerbestrasse 11, 6330 Cham, Switzerland

For
June Coleman Magrab

Preface

This book presents a concise and focused introduction to engineering statistics, emphasizing topics and concepts that a practicing engineer is mostly likely to use: the display of data, confidence intervals, hypothesis testing, fitting straight lines to data, and designing experiments to find the impact of process changes on a system or its output. The book introduces the language of statistics, contains sufficient detail so that there is no mystery as to how equations come about, makes extensive use of tables to collect and summarize in one place important formulas and concepts, and utilizes enhanced graphics that are packed with visual information to illustrate the meaning of the equations and their usage.

The book can be used as an introduction to the subject, to refresh one's knowledge of engineering statistics, to complement course materials, as a study guide, and to provide a resource for laboratories where data acquisition and analysis are performed.

The book recognizes that there are many computational tools available to perform the statistical analysis of data, and, therefore, the numerical details in the more computationally intensive examples are omitted. However, to facilitate all numerical calculations used in the book and to permit one to further explore the material, a set of interactive graphics (IGs) has been created using Mathematica®. The IGs are designed to do specific tasks in the context of the material. The IGs can be used to expand many of the figures by changing the figure's parameters, to evaluate relations in tables, to verify examples, and to solve exercises. One IG determines values for the probability and inverse probability of the normal, t, chi square, and f distributions. They do not require any programming, and their use is self-explanatory in the context of the material. The IGs are on the publisher's website as are their description and how to acquire the free software to run them.

The book consists of four chapters. In each chapter, there are numerical examples to illustrate the application of the results and exercises at the end of the chapter. Each example cites the applicable IG that can be used to verify the results and in the more computationally intensive ones to evaluate and plot them. The tables of data that accompany several of the exercises are available in a text file on the publisher's website and formatted so that they can be copied into any analysis software program that is used. A solution manual is also available on the publisher's website.

In Chap. 1, we introduce several fundamental aspects of statistics: its language, quantities used to describe data, visualization of data, discrete probability distributions, and terms that describe measurements. For the language of statistics, we define such terms as experiments, random samples, bias, and probability. For the quantities that describe data, we define the mean, median, mode, quartiles, expected value, unbiased variance, and covariance. For visualizing data, we discuss histograms, box-whisker plots, and scatter plots. Discrete probability distributions, the cumulative frequency function, the probability mass function, and the binomial distribution and the Poisson distribution are defined. Then, the concept of independent random variables and its significance is discussed. Lastly, terms used to describe measurements are defined and illustrated: accuracy, precision, repeatability, reproducibility, and stability.

In Chap. 2, we introduce continuous probability density functions: normal, lognormal, chi square, student t, f distribution, and Weibull. These probability density functions are then used to obtain the confidence intervals at a specified confidence level for the mean, differences in means, variance, ratio of variances, and difference in means for paired samples. These results are then extended to hypothesis testing where the p-value is introduced and the type I and type II errors are defined. The use of operating characteristic (OC) curves to determine the magnitude of these errors is illustrated. Also introduced is a procedure to obtain probability plots for the normal distribution as a visual means to confirm the normality assumption for data.

In Chap. 3, we provide derivations of the applicable formulas for simple and multiple linear regression. In obtaining these results, the partitioning of the data using the sum of squares identities and the analysis of variance (ANOVA) is introduced. The confidence intervals of the model's parameters are determined, and how an analysis of the residuals is used to confirm that the model is appropriate. Prior to obtaining the analytical results, we discuss why it is necessary to plot data before attempting to model it, state general guidelines and limits of a straight-line model to data, and show how one determines whether plotted data that appear to be nonlinear could be intrinsically linear. Hypothesis tests are introduced to determine which parameters in the model are the most influential.

In Chap. 4, the terms used in experimental design are introduced: response variable, factor, extraneous variable, level, treatment, blocking variable, replication, contrasts, and effects. The relations needed to analyze a one-factor experiment, a randomized complete block design, a two-factor experiment, and a 2^k-factorial experiment are derived. For these experiments, an analysis of variance is used to determine the factors that are most influential in determining its output and, when appropriate, whether the factors interact.

College Park, MD, USA Edward B. Magrab

Contents

Chapter 1
Descriptive Statistics and Discrete Probability Distributions

In this chapter, we introduce several fundamental aspects of statistics: its language, quantities used to describe data, visualization of data, discrete probability distributions, and terms that describe measurements. For the language of statistics, we define such terms as experiments, random samples, bias, and probability. For the quantities that describe data, we define the mean, median, mode, quartiles, expected value, unbiased variance, and covariance. For visualizing data, we discuss histograms, box-whisker plots, and scatter plots. Discrete probability distributions, the cumulative frequency function, the probability mass function, and the binomial distribution and the Poisson distribution are defined. Then, the concept of independent random variables and its significance is discussed. Lastly, terms used to describe measurements are defined and illustrated: accuracy, precision, repeatability, reproducibility, and stability.

1.1 Introduction

The world is awash with data, but data by themselves are often meaningless. It is statistics that provides the ability to give meaning to data and allows one to draw conclusions and to make decisions. One important aspect of statistics is that is can be used to quantify and interpret variability, which can then be used to identify meaningful changes in processes and experiments.

The approach to statistics used here will be analogous to that used to analyze many engineering problems; that is, based on a set of factors one chooses an appropriate model to describe the system, process, or artifact under consideration. These factors are usually determined from a combination of prior experience, engineering judgment, and direct measurements. In statistics, the vocabulary is different, but the procedure is

Supplementary Information The online version contains supplementary material available at [https://doi.org/10.1007/978-3-031-05010-7_1].

E. B. Magrab, *Engineering Statistics*, https://doi.org/10.1007/978-3-031-05010-7_1

similar: from a collection of measurements, one selects an appropriate statistical model to describe them and then uses this model to draw conclusions.

Many of the numerical values used in this and subsequent chapters are given without context. They are values that are employed to illustrate the various concepts and to illustrate the statistical inferences that can be drawn from their analysis. However, except for possibly their magnitudes, they could be representing such physical measurements as component/product time to failure, percentage shrinkage, number of defective components, variation in composition, production rates, reaction times, dimensional characteristics, weight, tensile/shear strength, displacement/velocity/acceleration, acoustic sound pressure level, flow rate, and temperature to name a few. The interpretation in the physical domain is simply a specific application of the topic that is being discussed.

1.2 Definitions

Samples are collections of observations from measuring a specific characteristic of a system, process, or artifact. Each observation is given a numerical real value (positive, negative, or zero) and this numerical value is considered a *random variable*. The samples themselves can be the entirety of all possible measurements, called the *population*, or they can be a subset of the population.

Using a sample of a population is never as good as using the population; however, a sample is used because: (a) timeliness of results is important; (b) some measured characteristics may require the destruction of the artifact; and (c) the population is too large.

Random variables are classified as either *discrete* or *continuous*. A discrete random variable is a value from an enumerable set. For example, the number of students in a class, the number of occupants in city, and the number of defective components. A continuous random variable has any value in a specified interval, finite or infinite. For example, the life of an appliance, automobile gas mileage, and atmospheric temperature and pressure. The classification of whether a variable is discrete or continuous determines how it is defined mathematically. In either case, one ends up choosing a probabilistic model that best describes the sample. When the range of a discrete random variable is very large, it may be more convenient to represent them as a continuous random variable. Conversely, there are situations where a continuous random variable can be converted to a discrete random variable by, for example, placing the data into three groups, where each sample's value is assigned to one of three values to indicate which group.

In statistics, an *experiment* is a procedure that when repeated in the same manner every time results in different outcomes, with each outcome resulting in a random variable.

Probability quantifies the likelihood that an outcome of an experiment will occur.

A *random sample* is one in which a sample selected from a collection of samples had an equal chance (probability) of being selected.

Bias is any sample selection procedure that consistently overestimates or underestimates the measured values.

1.3 Statistical Measures and the Display of Data

We shall introduce several measures that can be used to describe a set of observations:

1. The relative frequency of the observations, which gives a sense of the probability of an observation having a specific value.
2. The mean, mode, and median, which give a sense of the location of the center of the data
3. The variance, which gives a sense of the variability or dispersion in the data about the location of the center of the data

We now define each of these quantities and their meaning for discrete random variables.

1.3.1 Histograms

Consider a set of samples S comprised of the random variables $x_j, j = 1, 2, \ldots, n$; that is, $S = \{x_1, x_2, x_3, \ldots, x_n,\}$. We create a set of m equal intervals $\Delta_i = (y_{i+1} - y_i)$, $i = 1, 2, \ldots, m$, where $\{y_1, y_2, \ldots, y_{m+1}\}$ are the end points of the intervals. The quantity Δ_i is often referred to as the ith bin. In creating these intervals, we require that $y_1 \leq x_{\text{min}}$ and $y_{m+1} > x_{\text{max}}$ where x_{max} is the maximum value in S and x_{min} is the minimum value in S. The *range* of the values in S is $x_{\text{max}} - x_{\text{min}}$. We denote the number of values in S that fall within Δ_i as n_i, where $y_i \leq X < y_{i+1}$ and X denotes a random variable from S. Then, the *relative frequency* p_i of S having a value in Δ_i is $p_i = n_i/n$. Therefore,

$$\sum_{i=1}^{m} p_i = 1 \tag{1.1}$$

When creating histograms, the quantity m is often selected as an integer close to $\sqrt{n}$ and that Δ_i is often selected as an integer.

The quantity p_i is an estimate of the probability that the elements of S will have a value in the interval Δ_i. To express this idea mathematically, we introduce the notation

$$P(y_i \leq X < y_{i+1}) = p_i \tag{1.2}$$

which reads: the probability that the random variable X will have a value between y_i and y_{i+1} is p_i. Then, the probability of finding a value in S that lies within the range of contiguous intervals Δ_j to Δ_k, which is the interval $\Delta_{kj} = (y_k - y_j)$, where $1 \leq j \leq m - 1$ and $j + 1 \leq k \leq m$, is

Table 1.1 Data used in the construction of a histogram ($n = 90$) with the minimum and maximum values identified in bold

503	487	499	668	411	390	322	918	925
585	752	436	451	184	953	656	548	559
869	1000	1005	592	830	265	707	818	444
453	616	409	437	346	**30**	792	621	592
574	860	336	607	637	794	527	832	739
322	802	306	495	640	719	482	578	201
289	171	351	1068	737	843	**1441**	270	560
618	470	370	439	858	305	570	709	514
629	441	404	991	758	212	638	521	572
250	522	494	448	550	766	480	792	272

$$P(y_j \le X < y_k) = \sum_{i=j}^{k} p_i$$

The above expression leads to the idea of the *cumulative probability* $F_k(x)$, which is given by

$$F_k(x) = P(X < y_k) = \sum_{i=1}^{k} p_i \tag{1.3}$$

Equation (1.3) gives the probability of having a value from S in the region $y_k - y_1$, and this probability is denoted $F_k(x)$. It is seen from Eq. (1.1) that $F_m(x) = 1$.

The *mode* occurs at that Δ_i where n_i (or p_i) is maximum. This interval is the one in which a sample from S has the highest probability of occurring.

To illustrate these definitions, consider the data shown in Table 1.1. These data are sorted so that the first element has the smallest value and the last element has the largest value; that is, we create the ordered set $\widetilde{S} = \{\widetilde{x}_1, \widetilde{x}_2, \widetilde{x}_3, \ldots, \widetilde{x}_n\}$, where $\widetilde{x}_{j+1} \ge \widetilde{x}_j \; j = 1, 2, \ldots, n-1$. After a little experimentation, it is decided to initially set $m = 8$ and $\Delta_i = 200$; therefore, $y_i = 200(i - 1)$, $i = 1, 2, \ldots, 9$. We shall present the data three ways.[1] The first way is shown in Fig. 1.1, which displays the actual values that appear in each bin and encases these values in a rectangular box. Underneath each ith box is the number of values that appear in that box denoted n_i, the corresponding fraction of the total values denoted p_i, and the sum of the fraction of values that appear in the present bin and all those preceding denoted $F_i(x)$.

The second way that these data are displayed is in the traditional way, which is that of a *histogram* shown in Fig. 1.2. In Fig. 1.2a, we have used $\Delta_i = 200$ and in Fig. 1.2b, $\Delta_i = 100$. In both figures, we have chosen to plot the relative frequency of the data in the bins and to place thevalue of n_i corresponding to that Δ_i above each bar. It is seen in Fig. 1.2b that the smaller bin interval provides a better indication of

[1] All symbolic work, numerical work, and graphics are obtained with Mathematica. Mathematica is a registered trademark of Wolfram Research, Inc.

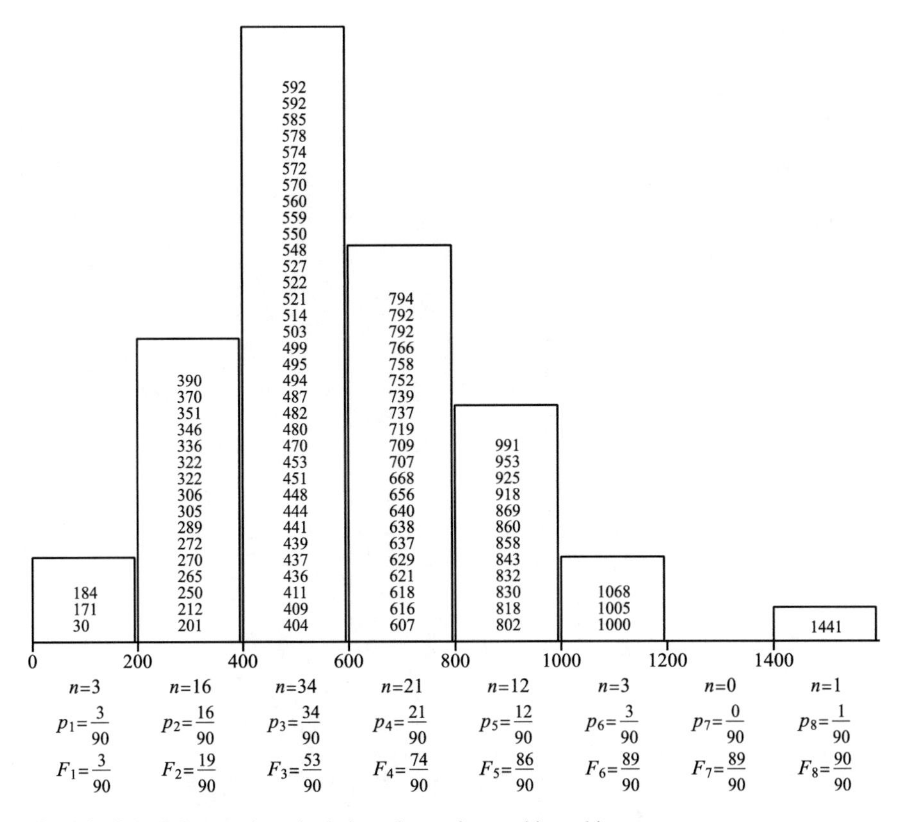

Fig. 1.1 Calculations and manipulations that go into making a histogram

the distribution of the data. As expected by the nature of the bin collection procedure, the number of values between 0 and 200 in Fig. 1.2a equals the sum of the number of values in the bins in the same interval in Fig. 1.2b. Similarly, for the bins between 200 and 400, and so on. The cumulative probability (or cumulative relative frequency) for the data presented in Fig. 1.2b is shown in Fig. 1.3. Another form of representing the data is to combine the histogram with a *scatter plot* of the data as shown in Fig. 1.4. The order of the samples in the scatter plot has no effect on the shape of the histogram.

Figures 1.2 and 1.3 can be replicated for any data set with interactive graphic IG1–1.

Histograms are used to visualize the distribution of the data and, as discussed subsequently, can be used to obtain an approximation to a quantity called the *probability mass function* and when continuous models are used, the *probability density function*. The shape of a histogram also provides some indication as to which probability model one should use to represent the data. Histograms additionally show the extent of the variability or 'spread' of the values about its central location, whether the data tend to be symmetrical or skewed, and whether the data are *bimodal*; that is, they have two separate and distinguishable groupings, each with

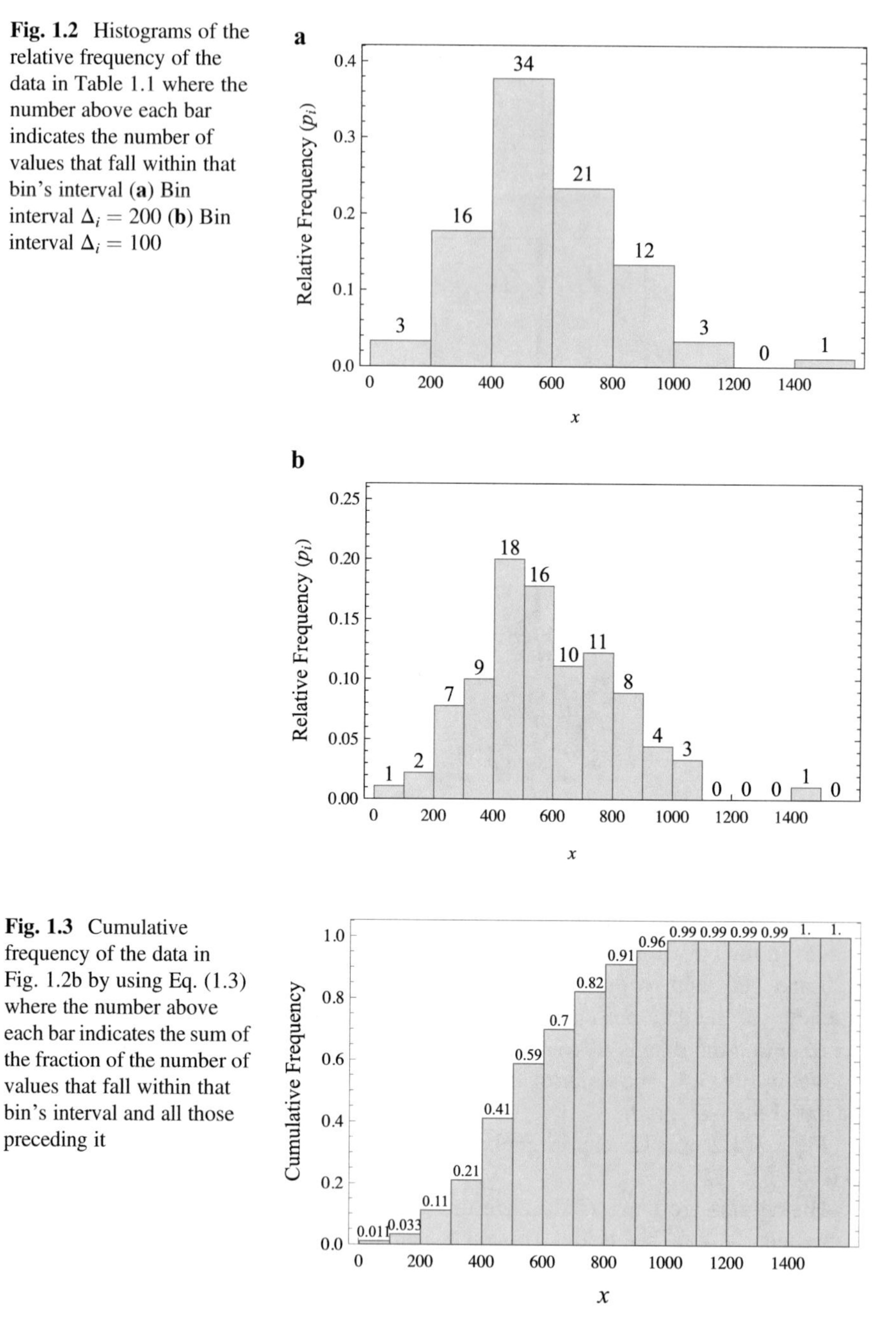

Fig. 1.2 Histograms of the relative frequency of the data in Table 1.1 where the number above each bar indicates the number of values that fall within that bin's interval (**a**) Bin interval $\Delta_i = 200$ (**b**) Bin interval $\Delta_i = 100$

Fig. 1.3 Cumulative frequency of the data in Fig. 1.2b by using Eq. (1.3) where the number above each bar indicates the sum of the fraction of the number of values that fall within that bin's interval and all those preceding it

its own peak. This latter case may occur, for example, when a physical characteristic of humans is measured, and this characteristic depends on gender. Other examples are when water/electrical usage depends on the season, or the measurement of travel time depends on the time of day and day of the week.

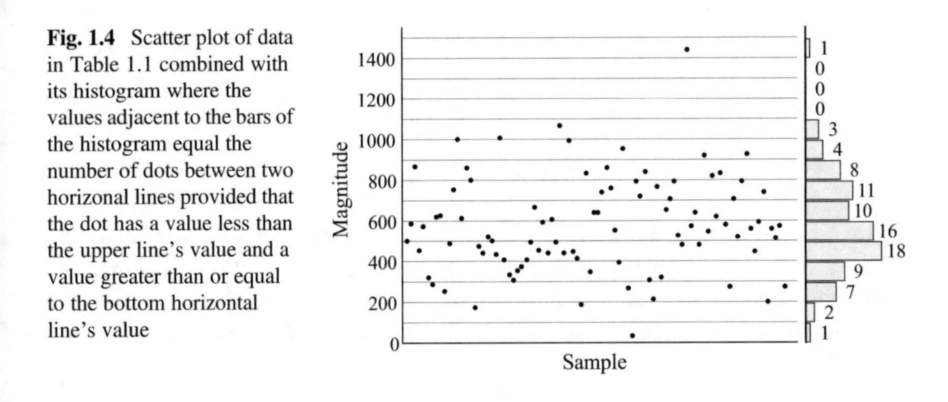

Fig. 1.4 Scatter plot of data in Table 1.1 combined with its histogram where the values adjacent to the bars of the histogram equal the number of dots between two horizonal lines provided that the dot has a value less than the upper line's value and a value greater than or equal to the bottom horizontal line's value

1.3.2 Sample Median and Quartiles

The *median* is that value for which the random variable is equally likely to be either smaller than or larger than this value. An advantage of the median is that it is not influenced by the magnitude of the values, just their number. The determination of the median depends on whether the number of values n in S is even or odd. Again, using the ordered set $\widetilde{S}$, for n odd the median $\widehat{\widetilde{x}}$ is the sample

$$\widehat{\widetilde{x}} = \widetilde{x}_{(n+1)/2} \tag{1.4}$$

and for n even, the median is the average value

$$\widehat{\widetilde{x}} = \frac{1}{2}\left(\widetilde{x}_{n/2} + \widetilde{x}_{n/2+1}\right) \tag{1.5}$$

When $\widetilde{S}$ is divided into four equal parts, the division points are called quartiles. The upper quartile defines the 75th percentile denoted $q_{0.75}$ for which 75% of the values are smaller than the division point, and the lower quartile defines the 25th percentile denoted $q_{0.25}$ for which 25% of the values are smaller than the division point. The 50th quartile denoted $q_{0.50}$ divides the data so that 50% of the values are smaller. The 50th quartile is the median value $\widehat{\widetilde{x}}$ of the sample given by either Eqs. (1.4) or (1.5). When $q_{0.75}$ - $q_{0.50}$ is substantially different from $q_{0.50}$ - $q_{0.25}$, the data are said to be *skewed*.

The determination of the 25th and 75th percentiles also depend on whether n is even or odd. When n is even, $q_{0.25}$ is obtained from the median of those values in $\widetilde{S}$ that are less than $\widehat{\widetilde{x}}$; that is, the values $\left\{\widetilde{x}_1, \widetilde{x}_2, \ldots, \widetilde{x}_{n/2}\right\}$ and $q_{0.75}$ is obtained from the median of those values in $\widetilde{S}$ that are greater than $\widehat{\widetilde{x}}$; that is, the values $\left\{\widetilde{x}_{n/2+1}, \widetilde{x}_{n/2+2}, \ldots, \widetilde{x}_n\right\}$. When n is odd, $q_{0.25}$ is obtained from the average of the median of each of the following sets of values: $\left\{\widetilde{x}_1, \widetilde{x}_2, \ldots, \widetilde{x}_{(n-1)/2}\right\}$ and $\left\{\widetilde{x}_1, \widetilde{x}_2, \ldots, \widetilde{x}_{(n+1)/2}\right\}$. When n is odd, $q_{0.75}$ is obtained from the average of the median of each of the following sets of values: $\{\tilde{x}_{(n+1)/2}, \tilde{x}_{(n+1)/2+1}, \ldots, \tilde{x}_n\}$ and $\{\tilde{x}_{(n+1)/2+1}, \tilde{x}_{(n+1)/2+2}, \ldots, \tilde{x}_n\}$.

A quantity called the *interquartile range*, IQR, is defined as

$$\text{IQR} = q_{0.75} - q_{0.25} \tag{1.6}$$

and is used to provide an indication of the spread of the sample values about the median. Note that one-half of the data values are within the IQR. The interquartile range and quartiles are used to define *outliers*, which are sample values that are very far removed from the boundaries defined by the IQR. Introducing the following quantities

$$q_{ol} = q_{0.25} - 1.5\,\text{IQR} \qquad q_{oh} = q_{0.75} + 1.5\,\text{IQR} \tag{1.7}$$

outliers are defined as those values that are less than q_{ol} or greater than q_{oh}. It is noted that if the maximum value of the data $x_{\max} \le q_{oh}$, then $x_{\max}$ is not an outlier. However, if there are outliers, then the largest one would be $x_{\max}$. If the minimum value of the data $x_{\min} \ge q_{oh}$, then $x_{\min}$ is not an outlier. However, if there are outliers, then the smallest one would be $x_{\min}$.

These various quartile quantities are used to construct a *box-whisker plot* (also denoted a box plot, box whisker diagram, and box whisker chart), which is shown in Fig. 1.5. The determination of the length of the whiskers is indicated in the figure.

The box-whisker plot is illustrated using the data in Table 1.1 and the results are shown in Fig. 1.6. From these data, it was found that $q_{0.25} = 436$, $q_{0.50} = 559.5$,

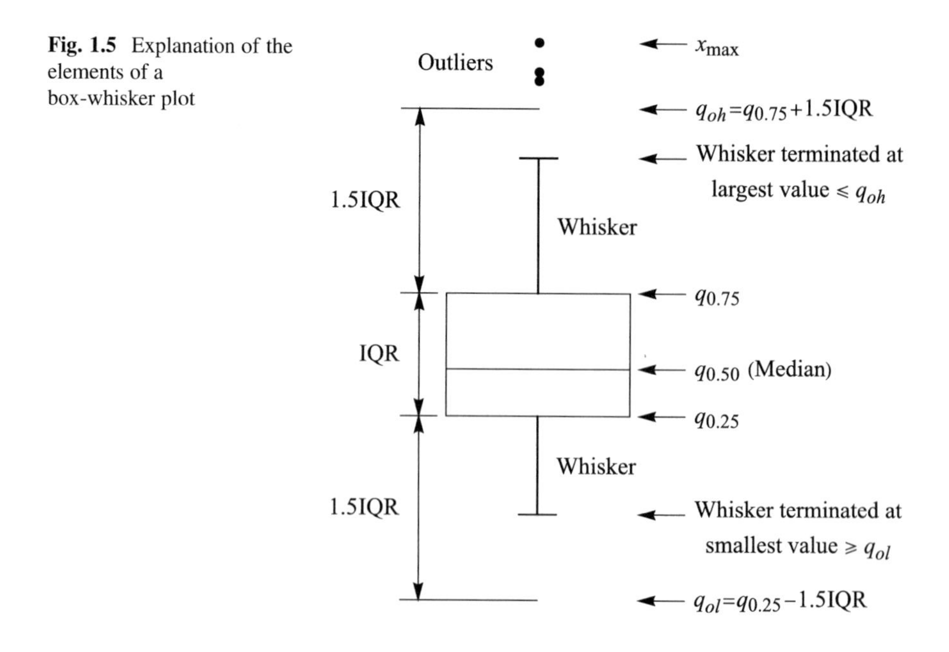

Fig. 1.5 Explanation of the elements of a box-whisker plot

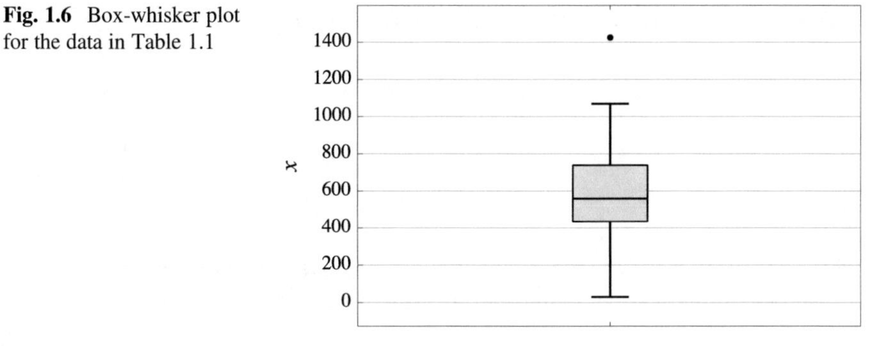

Fig. 1.6 Box-whisker plot for the data in Table 1.1

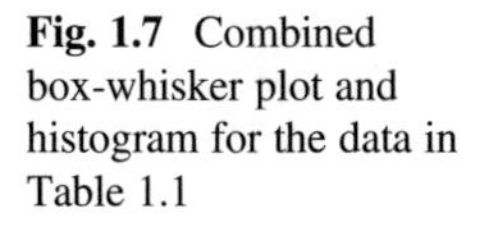

Fig. 1.7 Combined box-whisker plot and histogram for the data in Table 1.1

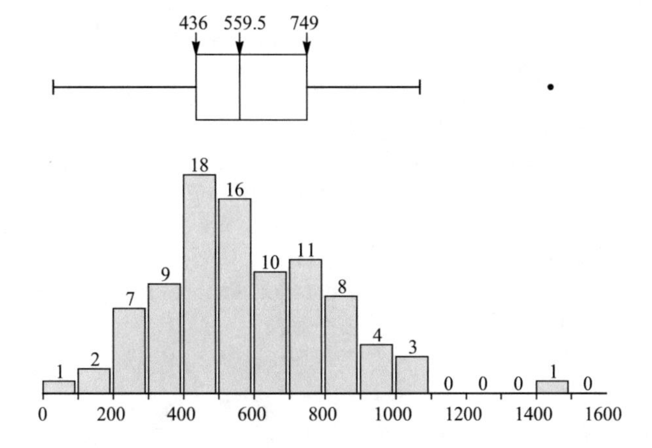

$q_{0.75} = 739$ and, therefore, IQR $= 303$, $q_{ol} = -18.5$, and $q_{oh} = 1193.5$. It is seen from Fig. 1.6 (and Fig. 1.2) that the data are skewed. This can be quantified by noting that $q_{0.75} - q_{0.50} = 179.5$ and $q_{0.50} - q_{0.25} = 123.5$. From Table 1.1, it is seen that $x_{\min} = 30$ and, therefore, there are no outliers of any of the smallest values and the lowest whisker terminates at $x_{\min}$. On the other hand, it is seen from Table 1.1 that there is one value that is greater than q_{oh} and that is $x_{\max} = 1441$; therefore, this value is an outlier. Also, the whisker for the largest values terminates from the ordered data at $x_{89} = 1068$ (refer to Fig. 1.1), since this value is less than $q_{oh} = 1193.5$ and the next largest value exceeds q_{oh}.

Figure 1.6 can be replicated for any data set with interactive graphics IG1–1 and for multiple data sets with IG1–2.

To place the box-whisker plot in context with its histogram, both have been plotted in Fig. 1.7 for the data in Table 1.1.

1.3.3 Sample Mean and the Expected Value

The mean of the values in S, denoted $\overline{x}$, is defined as

$$\overline{x} = \frac{1}{n} \sum_{i=1}^{n} x_i \tag{1.8}$$

where n is the number of independent samples that are used to determine the mean. The number of independent samples is often referred to as the number of *degrees of freedom*.

If each value in S has a constant value added to it, say b, and each value in S is multiplied by the same quantity, say a; that is, $x_i \to b + ax_i$, then the mean becomes

$$\overline{x} \to \frac{1}{n} \sum_{i=1}^{n} (b + ax_i) = \frac{1}{n} \sum_{i=1}^{n} b + \frac{a}{n} \sum_{i=1}^{n} x_i = b + a\overline{x} \tag{1.9}$$

where $\overline{x}$ is given by Eq. (1.8).

Subsequently, we shall be considering the mean of two different sets of observations and our notation will vary accordingly. When we are considering a sample, the mean will be denoted $\overline{x}$ and when we are considering a population, the mean will be denoted μ.

We now introduce the notation for a quantity called the *expected value* of X, which is defined as

$$E(X) = \sum_{i=1}^{n} x_i p_i \tag{1.10}$$

where $p_i = P(X = x_i)$ is the probability of each data value appearing in the sample. The quantity $P(X = x_i)$ is called the *probability mass function* of a discrete random variable. Equation (1.10) is the sum of the weighted values of x_i, where p_i are the weights. The expected value $E(X)$ gives the mean value.

If we use the results from a histogram, then for a selected bin Δ_i we assume that it is equally likely that X can have any value in Δ_i. We assume that the midpoint of Δ_i is a reasonable representation of each of these values. Then, corresponding to p_i there is the value $x_i = (i - 1)\Delta_i + \Delta_i/2$, $i = 1, 2, \ldots, m$, and Eq. (1.10) can be written as

$$E(X) = \sum_{i=1}^{m} p_i (i - 1/2)\Delta_i = \sum_{j=1}^{m} \frac{n_i}{n} (i - 1/2)\Delta_i \tag{1.11}$$

We illustrate this notation by using the data in Table 1.1 and the results shown in Fig. 1.2. Then, using Eq. (1.8), we find that $\overline{x} = 576.08$. Next, we use the n_i given in Fig. 1.2a, which were obtained for $\Delta_i = 200$ and $m = 8$, and from Eq. (1.11) find that

$E(X) = 582.22$. If we use the data shown in Fig. 1.2b in which $\Delta_i = 100$ and $m = 16$, then from Eq. (1.11) we find that $E(X) = 578.89$. As the bin size decreases and m increases proportionally, the expected value determined from the histogram approaches the mean computed from Eq. (1.8).

1.3.4 Sample Variance

The *unbiased* estimate of the variance of the values in S, denoted s^2, is called the *sample variance* and is defined as

$$s^2 = \frac{1}{n-1}\sum_{i=1}^{n}(x_i - \bar{x})^2 = \frac{S_{xx}}{n-1} = \frac{1}{n-1}\left[\sum_{i=1}^{n}x_i^2 - n\bar{x}^2\right] \tag{1.12}$$

where

$$S_{xx} = \sum_{i=1}^{n}(x_i - \bar{x})^2 = \sum_{i=1}^{n}x_i^2 - n\bar{x}^2 \tag{1.13}$$

is called the *sum of squares* and has been introduced in anticipation of subsequent discussions concerning regression analysis and the analysis of variance. The assertion that Eq. (1.12) is an unbiased estimate of the variance is proven in Sect. 1.3.7. While two sets of samples can have the same mean, one cannot assume that their variances are also equal.

The *standard deviation*, denoted s, is the positive square root of the variance.

If all the values in S are equal, then $s^2 = 0$. If each value in S is multiplied by a constant, say a, and if each value in S has a constant value added to it, say b; that is, $x_i \to b + ax_i$, then the variance becomes

$$\begin{aligned} s^2 &\to \frac{1}{n-1}\sum_{i=1}^{n}(b + ax_i - \bar{x})^2 = \frac{1}{n-1}\sum_{i=1}^{n}(b + ax_i - (b + a\bar{x}))^2 \\ &\to \frac{a^2}{n-1}\sum_{i=1}^{n}(x_i - \bar{x})^2 = a^2 s^2 \end{aligned} \tag{1.14}$$

where s^2 is given by Eq. (1.12), $\bar{x}$ is given by Eq. (1.8), and we have used Eq. (1.9).

The sample variance plays an important role in drawing inferences about the population variance and the sample mean and sample standard deviation are used to draw inferences about the population mean.

Subsequently, we shall be considering the variances of two different sets of observations and our notation will vary accordingly. When we are considering a sample, the variance will be denoted s^2 and when we are considering a population, the variance will be denoted σ^2 and in its computation μ is used.

We define the variance of a random variable Var(X) as the expected value of the $(X - E(X))^2$; thus,

$$
\begin{aligned}
\mathrm{Var}(X) &= E\left((X - E(X))^2\right) = \sum_{i=1}^{n} (x_i - E(X))^2 p_i \\
&= \sum_{i=1}^{n} x_i^2 p_i - 2E(X) \sum_{i=1}^{n} x_i p_i + [E(X)]^2 \sum_{i=1}^{n} p_i \\
&= E\left(X^2\right) - [E(X)]^2
\end{aligned}
\tag{1.15}
$$

where we have used Eqs. (1.1) and (1.10). The quantity

$$
E\left(X^2\right) = \sum_{i=1}^{n} p_i x_i^2 \tag{1.16}
$$

is the expected value of X^2. Then, from Eq. (1.15),

$$
E\left(X^2\right) = \mathrm{Var}(X) + [E(X)]^2 \tag{1.17}
$$

The definition of variance is extended to a quantity called the *covariance*, which is defined as

$$
\begin{aligned}
\mathrm{Cov}(X, Y) &= E((X - E(X))(Y - E(Y))) \\
&= E(XY) - E(X)E(Y)
\end{aligned}
\tag{1.18}
$$

where

$$
E(XY) = \sum_{j=1,}^{m} \sum_{i=1}^{n} x_i y_j p_{ij} \tag{1.19}
$$

and $p_{ij} = P(X = x_i,\ Y = y_j)$ is the *joint probability function*, $p_{ij} > 0$, and

$$
\sum_{j=1}^{m} \sum_{i=1}^{n} p_{ij} = 1
$$

It is seen from Eqs. (1.18) and (1.15) that Cov(X,X) = Var(X).

When X and Y are *independent*, $p_{ij} = P(X = x_i)P(Y = y_j)$ and Eq. (1.19) becomes

$$
E(XY) = \sum_{i=1}^{n} x_i p_i \sum_{j=1}^{m} y_j p_j = E(X)E(Y) \tag{1.20}
$$

and, therefore, from Eq. (1.18), we see that when X and Y are *independent* $\text{Cov}(X,Y) = 0$.

Using Eq. (1.18), it is straightforward to show that

$$\text{Cov}(X + Z, Y) = \text{Cov}(X, Y) + \text{Cov}(Z, Y) \tag{1.21}$$

1.3.5 Probability Mass Function

We now take the definitions given by Eqs. (1.1), (1.3), (1.10) and (1.15) and formalize them with the following notational changes. The *probability mass function* denoted $f(x)$ for a set of n discrete random variables is

$$f(x_i) = P(X = x_i) \qquad i = 1, 2, .., n \tag{1.22}$$

where X can have any value x_i from the set of samples, $f(x_i) \geq 0$, and

$$\sum_{i=1}^{n} f(x_i) = 1 \tag{1.23}$$

The *cumulative distribution function* $F(x)$, where $0 \leq F(x) \leq 1$, is

$$F(x) = P(X \leq x) = \sum_{x_i \leq x} f(x_i) \tag{1.24}$$

The expected value of X is

$$E(X) = \sum_{x} xf(x) = \overline{x} \tag{1.25}$$

and the variance of X is

$$\text{Var}(X) = E\left((X - \overline{x})^2\right) = \sum_{x} (x - \overline{x})^2 f(x) = \sum_{x} x^2 f(x) - \overline{x}^2 \tag{1.26}$$

It is possible for two samples to have the same mean and variance, but different probability mass functions.

1.3.6 Independent Random Variables

We now discuss the significance of *independent random variables*. Consider two processes from which we take samples. The samples from one process are denoted X and those from the other process Y. These random variables are *independent* if the value X does not depend on the value Y. In other words, the process that created X is unrelated to that which created Y. The assumption of the output of one variable being statistically independent from another variable is frequently made and often can be justified with a physical basis.

The consequences of independent random variables are as follows. Consider the two independent random variables X and γY, where $\gamma = \pm 1$ and $\gamma^2 = 1$. The expected value of their sum is

$$E(X + \gamma Y) = E(X) + \gamma E(Y)$$

and their variance is

$$\begin{aligned}
\mathrm{Var}(X + \gamma Y) &= E\left[(X + \gamma Y)^2\right] - [E(X + \gamma Y)]^2 \\
&= E\left(X^2\right) + E\left(Y^2\right) + 2\gamma E(XY) \\
&\quad -[E(X)]^2 - [E(Y)]^2 - 2\gamma E(X)E(Y) \\
&= E\left(X^2\right) - [E(X)]^2 + E\left(Y^2\right) - [E(Y)]^2 \\
&\quad +2\gamma(E(XY) - E(X)E(Y)) \\
&= \mathrm{Var}(X) + \mathrm{Var}(Y) + 2\gamma \mathrm{Cov}(X, Y)
\end{aligned} \tag{1.27}$$

However, when X and Y are independent, we have showed by using Eq. (1.20) that $\mathrm{Cov}(X,Y) = 0$. Therefore, Eq. (1.27) reduces to

$$\mathrm{Var}(X + \gamma Y) = \mathrm{Var}(X) + \mathrm{Var}(Y) \tag{1.28}$$

It is seen from Eq. (1.28) that the variability of adding (or subtracting) two independent random variables always results in an *increase* in variability.

We now extend this result and consider the linear combination of n *independent* random variables X_i

$$Y = \sum_{i=1}^{n} a_i x_i + b \tag{1.29}$$

where a_i and b are constants. Then, the expected value of Y is

$$E(Y) = \sum_{i=1}^{n} a_i E(x_i) + b \tag{1.30}$$

The variance of Y is

$$\text{Var}(Y) = \text{Var}\left(\sum_{i=1}^{n} a_i x_i + b\right) = \sum_{i=1}^{n} a_i^2 \text{Var}(X_i) \tag{1.31}$$

where we have used Eqs. (1.28) and (1.14) and the fact that Var(b) = 0. In subsequent chapters, we shall be using frequently the fact that Var(aX) = a^2Var(X).

We now assume that we have a collection of n independent random samples $S = \{x_1, x_2, \ldots, x_n\}$ with each sample having an expected value of $\mu = E(X_i)$ and a variance $\sigma^2 = \text{Var}(X_i)$. This type of assumption is usually denoted by the abbreviation i.i.d. (or iid) to indicate that they are *independent and identically distributed.* Then, from Eq. (1.29) with $a_i = 1/n$ and $b = 0$, we see that

$$Y = \frac{1}{n}\sum_{i=1}^{n} x_i = \overline{X} \tag{1.32}$$

Then, the expected value of Y is

$$E(Y) = E\left(\overline{X}\right) = \frac{1}{n}\sum_{i=1}^{n} E(X_i) = \frac{1}{n}\sum_{i=1}^{n} \mu = \mu \tag{1.33}$$

The variance is obtained by using Eq. (1.31) with $a_i = 1/n$ and (1.32) to arrive at

$$\text{Var}(Y) = \text{Var}\left(\overline{X}\right) = \sum_{i=1}^{n} a_i^2 \text{Var}(X_i) = \frac{1}{n^2}\sum_{i=1}^{n} \sigma^2 = \frac{\sigma^2}{n} \tag{1.34}$$

which indicates that as the number of samples used to calculate the mean increases, the variability about the mean decreases.

We note from Eqs. (1.15), with a change of notation, that

$$\text{Var}(Y) = E\left(Y^2\right) - [E(Y)]^2$$

Then, Eqs. (1.33) and (1.34) give

$$E(\overline{X}^2) = \text{Var}\left(\overline{X}\right) + \left[E\left(\overline{X}\right)\right]^2 = \frac{\sigma^2}{n} + \mu^2 \tag{1.35}$$

1.3.7 Unbiased Variance

We return to Eq. (1.12) and explain the division by $n - 1$. Before doing so, we introduce the definition of *bias*. If the estimate of a statistic is $\widehat{\theta}$ and the true (population) value is θ, then $\widehat{\theta}$ is an unbiased estimate of θ when

$$E(\widehat{\theta}) = \theta \tag{1.36}$$

If the statistic is biased, then the bias is $B(\widehat{\theta}) = E(\widehat{\theta}) - \theta \neq 0$. For example, if $\widehat{\theta} = \overline{X}$ and $\theta = \mu$, we see from Eq. (1.33) that $E(\overline{X}) = \mu$ and, therefore, $E(\widehat{\theta}) - \theta = 0$ and $\overline{X}$ is an unbiased estimate of μ. In what follows, we are going to determine under what conditions $E(\widehat{s^2}) = \sigma^2$, where $\widehat{s^2}$

$$\widehat{s^2} = \frac{1}{n}\sum_{i=1}^{n}(x_i - \overline{x})^2 \tag{1.37}$$

We assume that the samples are independent and identically distributed such that $\mu = E(X_i)$ and $\sigma^2 = \mathrm{Var}(X_i) = E[(X_i - \mu)^2]$. Next, we consider the identity

$$\begin{aligned}\sum_{i=1}^{n}(x_i - \mu)^2 &= \sum_{i=1}^{n}((x_i - \overline{x}) + (\overline{x} - \mu))^2 \\ &= \sum_{i=1}^{n}(x_i - \overline{x})^2 + 2(\overline{x} - \mu)\sum_{i=1}^{n}(x_i - \overline{x}) + \sum_{i=1}^{n}(\overline{x} - \mu)^2 \\ &= \sum_{i=1}^{n}(x_i - \overline{x})^2 + n(\overline{x} - \mu)^2\end{aligned} \tag{1.38}$$

since, from Eq. (1.8),

$$\sum_{i=1}^{n}(x_i - \overline{x}) = \sum_{i=1}^{n}x_i - n\overline{x} = \sum_{i=1}^{n}x_i - \sum_{i=1}^{n}x_i = 0$$

Rearranging Eq. (1.38) as

$$\sum_{i=1}^{n}(x_i - \overline{x})^2 = \sum_{i=1}^{n}(x_i - \mu)^2 - n(\overline{x} - \mu)^2$$

the expected value of the sample variance $\widehat{s^2}$ is

$$\begin{aligned}
E\left[\widehat{s^2}\right] &= E\left[\frac{1}{n}\sum_{i=1}^{n}(x_i-\overline{x})^2\right] = E\left[\frac{1}{n}\sum_{i=1}^{n}(x_i-\mu)^2-(\overline{x}-\mu)^2\right] \\
&= \frac{1}{n}\sum_{i=1}^{n}E\left[(x_i-\mu)^2\right]-E\left[\overline{x}^2-2\overline{x}\mu+\mu^2\right] \\
&= \frac{1}{n}\sum_{i=1}^{n}\mathrm{Var}(X_i)-E\left[\overline{x}^2\right]+2\mu E[\overline{x}]-\mu^2 \\
&= \frac{1}{n}\left(n\sigma^2\right)-E\left[\overline{x}^2\right]+\mu^2=\sigma^2-\left(\frac{\sigma^2}{n}+\mu^2\right)+\mu^2 \\
&= \frac{n-1}{n}\sigma^2
\end{aligned} \tag{1.39}$$

Thus, the expected value of $\widehat{s^2}$ is biased by an amount $-\sigma^2/n$ and, therefore, the variance determined by Eq. (1.37) underestimates the true variance σ^2. In order to use an estimate of the population variance σ^2, we instead consider the expected value $E\left[n\widehat{s^2}/(n-1)\right]$, which equals σ^2. Hence, we use

$$s^2=\frac{n\widehat{s^2}}{n-1}=\frac{1}{n-1}\sum_{i=1}^{n}(x_i-\overline{x})^2 \tag{1.40}$$

which is an unbiased estimate of the population variance σ^2; that is, $E[s^2]=\sigma^2$.

1.4 Binomial and Poisson Distributions

We shall consider two probability mass distributions: the binomial distribution and the Poisson distribution.

1.4.1 Binomial Distribution

The binomial distribution determines the probability of having exactly x successes in n independent trials (attempts) with each attempt having the constant probability p of success. The quantity $n - x$ is the number of "failures." The term "success" is context dependent. The binomial distribution is given by

$$f_b(x)=P(X=x)=\binom{n}{x}p^x(1-p)^{n-x} \qquad x=0,1,2,\ldots,n \tag{1.41}$$

where

$$\binom{n}{x} = \frac{n!}{x!(n-x)!} \qquad x = 0, 1, 2, \ldots, n \tag{1.42}$$

is the number of combinations of n distinct objects taken x at a time and, by definition, $0! = 1$. Equation (1.41) is shown in Fig. 1.8 for several combinations of p and n.

From the binomial expansion

$$(a+b)^n = \sum_{k=0}^{n} \binom{n}{k} a^k b^{n-k}$$

we see that with $a = p$ and $b = 1 - p$

$$1 = \sum_{k=0}^{n} \binom{n}{x} p^x (1-p)^{n-x} = \sum_{k=0}^{n} f_b(x) \qquad x = 0, 1, 2, \ldots, n \tag{1.43}$$

Therefore,

$$\sum_{k=m}^{n} f_b(x) = 1 - \sum_{k=0}^{m-1} f_b(x) \qquad x = 0, 1, 2, \ldots, n \tag{1.44}$$

It is noted that

$$\binom{n}{0} = \frac{n!}{0!(n-0)!} = 1 \quad \text{and} \quad P(X=0) = \binom{n}{0} p^0 (1-p)^n = (1-p)^n$$
$$\binom{n}{n} = \frac{n!}{n!(n-n)!} = 1 \quad \text{and} \quad P(X=n) = \binom{n}{n} p^n (1-p)^0 = p^n$$

The probability of having at least m $(0 \le m \le n)$ successes in n attempts is

$$P(X \ge m) = \sum_{x=m}^{n} \binom{n}{x} p^x (1-p)^{n-x} \tag{1.45}$$

and the probability of having no more than m $(0 \le m \le n)$ successes is

$$P(X \le m) = \sum_{x=0}^{m} \binom{n}{x} p^x (1-p)^{n-x} \tag{1.46}$$

The mean and variance, respectively, for the binomial distribution are

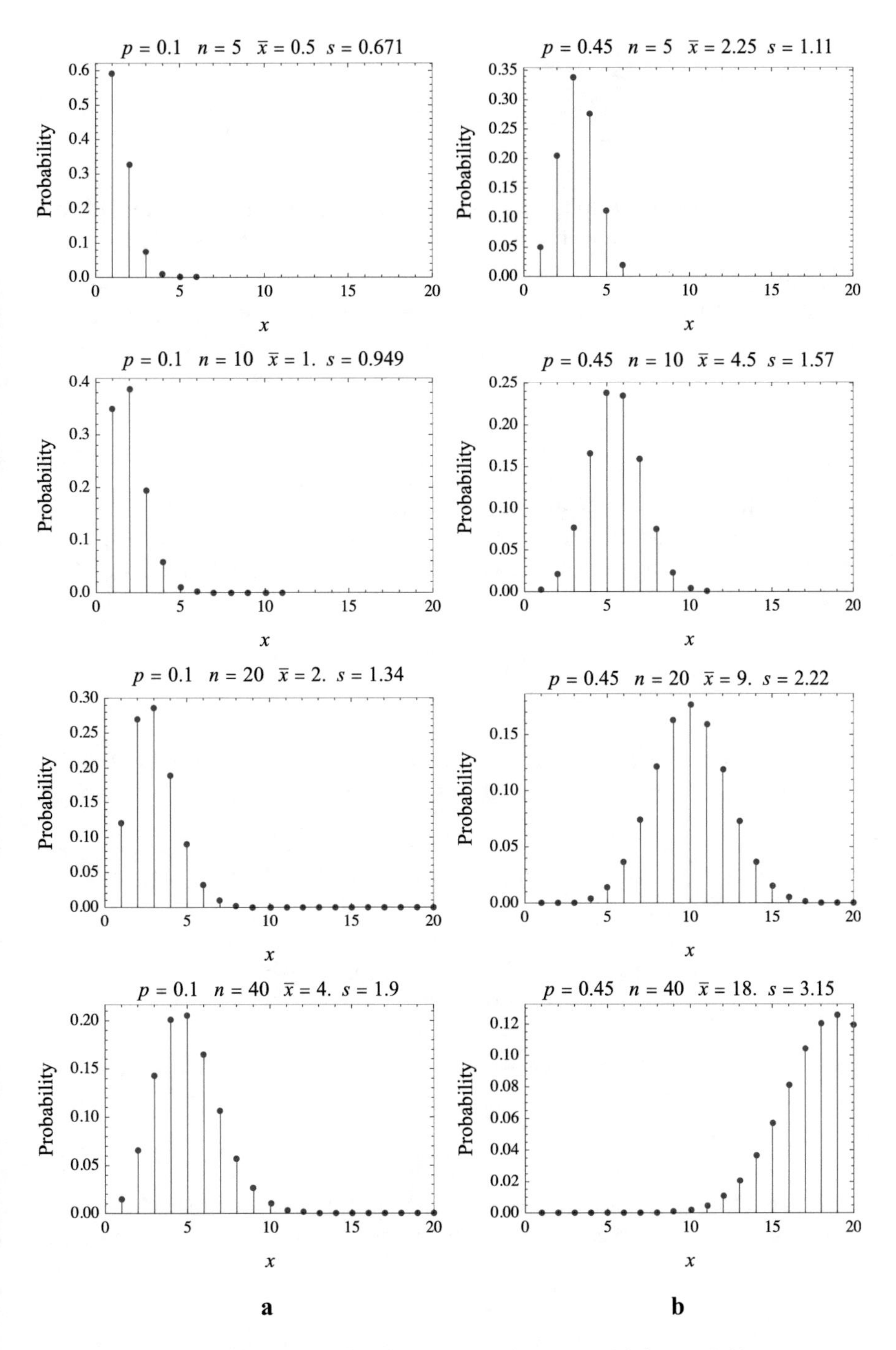

Fig. 1.8 Binomial distribution as a function of p, n, and x (**a**) $p = 0.1$ (**b**) $p = 0.45$

$$\begin{aligned} \bar{x} &= \sum_x x f_b(x) = np \\ s^2 &= \sum_x (x - \bar{x})^2 f_b(x) = np(1 - p) = \bar{x}(1 - p) \end{aligned} \tag{1.47}$$

We now consider the following example to illustrate the use of the binomial distribution. The values obtained in this example can be verified with IG1–3.

Example 1.1
(a) We shall determine the probability of getting exactly 3 'heads' when a coin is tossed 12 times. In this case, $n = 12$, $p = 0.5$, and $X = 3$. Then, Eq. (1.41) gives

$$P(X = 3) = \binom{12}{3} 0.5^3 (1 - 0.5)^{12-3} = 0.0537$$

(b) We shall determine the probability of a contaminant being present in exactly 2 of 15 samples of a product when there is a 12% chance of this contaminant being present in any one sample. In this case, $n = 15$, $p = 0.12$, and $X = 2$. Then, Eq. (1.41) gives

$$P(X = 2) = \binom{15}{2} 0.12^2 (1 - 0.12)^{15-2} = 0.2869$$

(c) In the situation given in (b) above, we shall determine the probability that 3 or more products are contaminated. In this case, we use Eqs. (1.44) and (1.45) to obtain

$$\begin{aligned} P(X \geq 3) &= \sum_{x=3}^{15} \binom{15}{x} 0.12^x (1 - 0.12)^{15-x} = 1 - \sum_{x=0}^{2} \binom{15}{x} 0.12^x (1 - 0.12)^{15-x} \\ &= 1 - \binom{15}{0} 0.12^0 (0.88)^{15} - \binom{15}{1} 0.12^1 (0.88)^{14} \\ &\quad - \binom{15}{2} 0.12^2 (0.88)^{13} \\ &= 0.265 \end{aligned}$$

(d) In the situation given in (b) above, we shall determine the probability that no more than 2 products are contaminated. In this case, we use Eq. (1.46) to obtain

$$
\begin{aligned}
P(X \leq 2) &= \sum_{x=0}^{2} \binom{15}{x} 0.12^x (1-0.12)^{15-x} \\
&= \binom{15}{0} 0.12^0 (0.88)^{15} + \binom{15}{1} 0.12^1 (0.88)^{14} + \binom{15}{2} 0.12^2 (0.88)^{13} \\
&= 0.735
\end{aligned}
$$

1.4.2 Poisson Distribution

The Poisson distribution estimates the probability of a random variable X having a certain number of outcomes during a given interval of time or over a specified length, area, or volume. It is assumed that X is distributed uniformly throughout the time interval or within the region of interest. If $\lambda > 0$, then the Poisson distribution is

$$
f_P(x) = P(X = x) = \frac{\lambda^x e^{-\lambda}}{x!} \qquad x = 0, 1, 2, \ldots, \tag{1.48}
$$

The expected value of X and its variance, respectively, are given by

$$
\begin{aligned}
E(X) &= \lambda \\
\text{Var}(X) &= \lambda
\end{aligned} \tag{1.49}
$$

Equation (1.48) is plotted in Fig. 1.9 for two values of λ.

In practice, $\lambda = \lambda_o \alpha$, where λ_o indicates the events per time, length, area, or volume and α states the corresponding duration, length, area, or volume over which λ_o is of interest. For example, if λ_o is the number of flaws per centimeter of material and we wish to determine the probability of N flaws in α centimeters, then $\lambda = \lambda_o \alpha$ is the non-dimensional quantity that is used.

It is noted that

$$
\sum_{x=0}^{\infty} P(X = x) = e^{-\lambda} \sum_{x=0}^{\infty} \frac{\lambda^x}{x!} = 1
$$

Then, the probability of having at least m events in the interval is

$$
P(X \geq m) = e^{-\lambda} \sum_{x=m}^{\infty} \frac{\lambda^x}{x!} = 1 - e^{-\lambda} \sum_{x=0}^{m-1} \frac{\lambda^x}{x!} \tag{1.50}
$$

and the probability of having no more than m events is

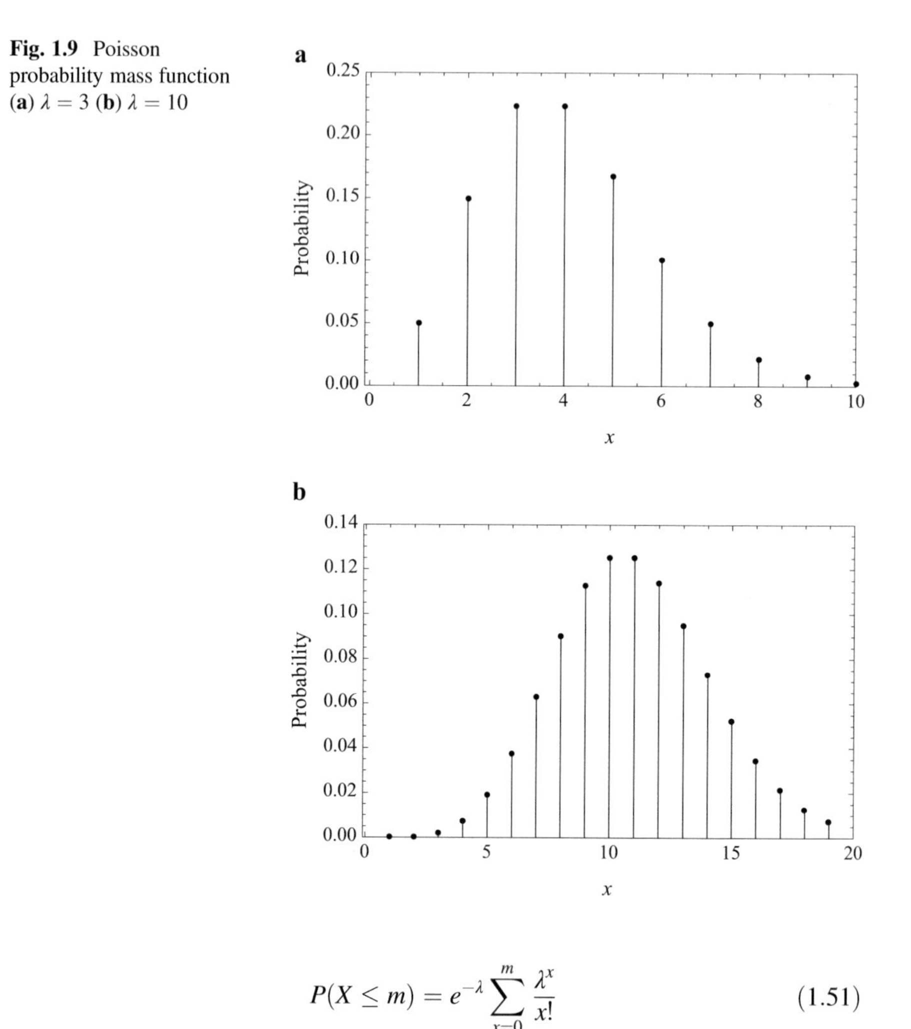

Fig. 1.9 Poisson probability mass function (**a**) $\lambda = 3$ (**b**) $\lambda = 10$

$$P(X \leq m) = e^{-\lambda} \sum_{x=0}^{m} \frac{\lambda^x}{x!} \tag{1.51}$$

The probability of having fewer than m events is

$$P(X < m) = e^{-\lambda} \sum_{x=0}^{m-1} \frac{\lambda^x}{x!} \tag{1.52}$$

The use of the binomial distribution is illustrated with the following example. The values obtained in this example can be verified with IG1–4.

Example 1.2
It is found that a glass fiber has on average 0.4 defects per meter.
(a) We shall determine the probability that this fiber could have exactly 12 defects in 25 meters. We find that $\lambda = 0.4 \times 25 = 10$. Then using Eq. (1.48)

$$P(X = 12) = \frac{10^{12}e^{-10}}{12!} = 0.0948$$

(b) We shall determine the probability that there are no more than 12 defects in 25 meters. In this case, we use Eq. (1.51) to obtain

$$P(X \leq 12) = e^{-10}\sum_{x=0}^{12}\frac{10^x}{x!} = 0.7916$$

(c) We shall determine the probability that there are no defects in 25 meters. We again use Eq. (1.48) and find that

$$P(X = 0) = \frac{10^0 e^{-10}}{0!} = 4.54 \times 10^{-5}$$

1.5 Definitions Regarding Measurements

All meaningful measurements are made by instruments traceable to United States standards or international standards. This traceability establishes the *accuracy* of the measuring devices. Hence, inaccuracy is the amount that measurements will on the average differ from the standard value. Regarding these measurements, the following terms are often used.

Precision The precision of a measurement is synonymous with the *resolution* of a measurement device's output; that is, the number of digits displayed or stored or the resolution of the analog display.

Repeatability The variation of the individual measurements of the same characteristic on the same artifact using the same measuring devices under the same conditions and location by the same person over a short period of time.

Reproducibility The variation on the average of measurements of the same characteristic taken by different persons using the same instruments under the same conditions on the same artifact.

Stability The variation on the average of at least two sets of measurements of the same characteristic taken by the same person using the same instruments under the same conditions on the same artifact but with each set taken at substantially different times.

These definitions are shown in Fig. 1.10.

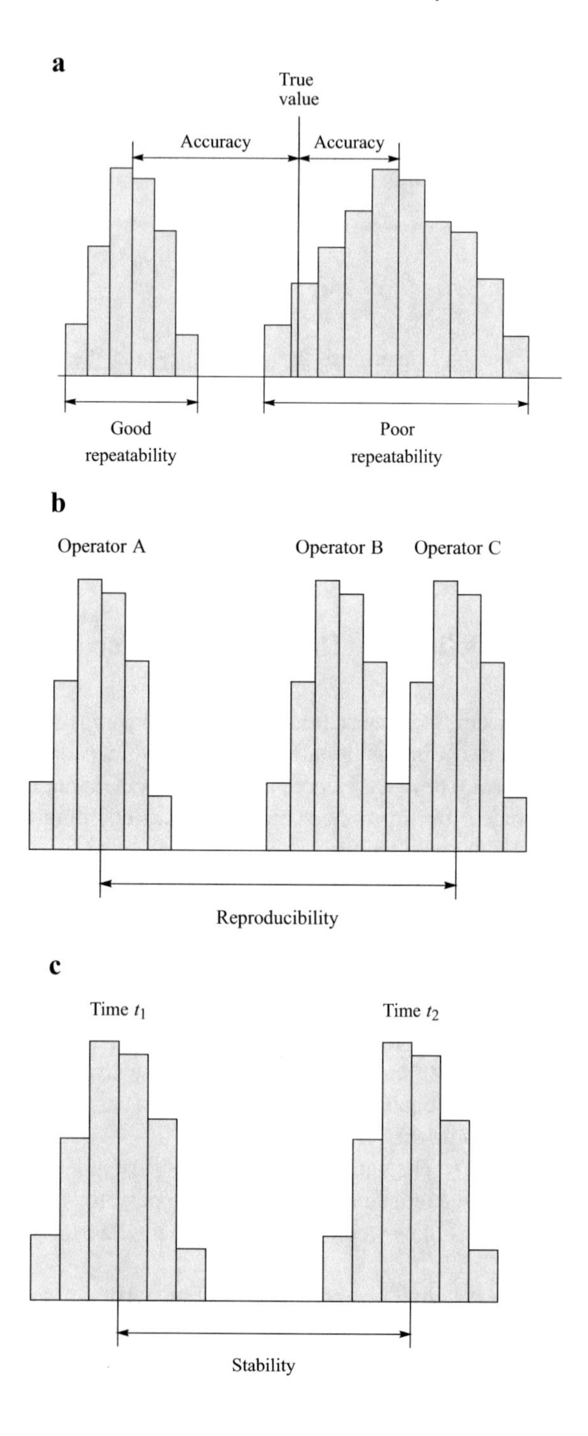

Fig. 1.10 Visualization of (**a**) Accuracy and repeatability (**b**) Reproducibility (**c**) Stability

Table 1.2 Data for Exercise 1.1

949	529	1272	311	583	512	505	461	765	367
1320	1307	1244	1164	562	1175	1338	1417	561	629
498	1093	590	1393	341	1723	1430	1595	326	577
309	1693	355	690	551	603	1487	716	590	764
1031	610	522	380	1326	430	414	492	574	509
945	235	808	630	1794	700	553	1680	1519	454
1274	572	305	376	1001	1593	1088	1004	1114	1132
1289	937	249	1271	1149	223	1334	1222	717	464
1561	656	797	1280	1318	1287	1438	1652	1609	1206
831	1149	1478	1072	1404	528	504	1506	772	1707
570	615	494	734	1237	1284	1423	1276	456	1180
533	1403	1486	1574	259	1654	1333	1824	1020	1343
1368	1363	691	1466	1464	798	873	764	645	1652
455	622	483	1616	904	1787	1875	1391	786	1563
396	823	1624	1695	1489	1242	1663	469	1880	1419
631	1131	355	987	432	1096	458	292	1362	340
1725	1354	426	467	1734	1200	879	2197	1414	
729	1469	1947	821	612	1438	1865	741	1454	

1.6 Exercises

Sections 1.3.1 and 1.3.2

1.1. For the data in Table 1.2:

(a) Determine the mean, standard deviation, median, mode, and the quantities necessary to create a box-whisker plot.
(b) Obtain a box-whisker plot.
(c) Obtain a histogram with bin width equal to 125.
(d) What conclusions can you draw from the histogram? Note the shortcomings of the box-whisker plot for these data.
(e) Compare the mean value $\pm$ one standard deviation from the mean with the 25% and 75% quartiles. Anything to be learned from this?

1.2. For the data in Table 1.3:

(a) Determine the mean, standard deviation, median, mode, and the quantities necessary to create a box-whisker plot.
(b) What can you infer from a comparison of the mean and the median?
(c) Obtain a box-whisker plot.
(d) Obtain a histogram with bin width equal to 100.
(e) What conclusions can you draw from the histogram?
(f) Does the box whisker plot offer any additional information that cannot be inferred from the histogram?
(g) Compare the mean value $\pm$ one standard deviation from the mean with the 25% and 75% quartiles. Anything to be learned from this?

Table 1.3 Data for Exercise 1.2

741	570	423	1011	104	562	147	484
385	444	250	603	404	304	412	732
775	948	411	517	241	463	490	556
408	610	365	708	455	352	516	874
343	352	172	934	844	141	234	962
632	371	557	555	377	332	255	887
786	289	379	220	452	662	252	115
135	515	987	281	660	225	662	528
306	580	530	354	226	358	623	
325	447	487	1082	353	350	390	

Table 1.4 Data for Exercise 1.3

Run 1	Run 2	Run 3	Run 4	Run 5
102	113	105	106	106
107	113	105	98	95
110	105	89	97	93
112	102	79	93	96
115	107	114	92	99
117	100	105	108	104
110	98	102	106	98
93	105	101	105	98
117	97	102	101	112
113	93	101	102	98
91	111	105	98	101
124	111	103	99	98
102	97	89	94	98
115	105	103	91	98
105	101	112	93	102
115	96	108	109	104
82	105	104	103	91
98	100	101	89	111
117	96	101	102	97
113	97	101	95	104

1.3. Five runs on the same experimental setup were made. The results are given in Table 1.4. Obtain a box-whisker plot of each set of data and plot them on the same graph. From an analysis of this graph, what conjectures can you make regarding such things as: bias, faulty experimental run, calibration (accuracy), stability, steady-state conditions, repeatability, and reproducibility.

Sect. 1.4.1

1.4. A group of 13 randomly selected items are weighed each day. The probability that this sample will have no item weighing less than the design weight is 0.15.

(a) What is the probability that in this group there are three items that weigh less than the design weight?

(b) What is the probability that in this group there are more than three items that weigh less than the design weight?
(c) In this group, what is the expected number of the items that will weigh less than the design weight.

1.5. It is known that 41% of components will survive a certain qualification test. If 15 components are tested:

(a) What is the probability that at least 10 will survive?
(b) What is the probability that 5 will survive?
(c) What is the probability that at least 3 but no more than 8 will survive?

1.6. A strength test of 20 samples has the probability of 0.246 that some of these samples will not meet a strength requirement.

(a) What is the expected number of samples that will not meet the requirement?
(b) What is the probability that seven samples will not meet this requirement?
(c) What is the probability that no more than three samples will not meet this requirement?

Sect. 1.4.2

1.7. The average number of defective products from a production line is 4.5 per day. If it can be assumed that the number of defective products follows a Poisson distribution, what is the probability that there will be:

(a) Fewer than 3 defective products on a given day.
(b) At least 30 defective products in a week.
(c) Exactly 4 defective products on a given day.
(d) Exactly 4 defective products in 3 of the next 5 days. Hint: Use the results of (c) above and Eq. (1.43).

1.8. The average number of parts that a production line makes per minute is 4.16. If the production line produces more than 8 parts/minute, the conveyor system will not be able to keep up. If it can be assumed that this process follows a Poisson distribution, then what is the probability that this critical production rate will be exceeded?

1.9. It is found that the probability of an accident on any given day is 0.0035 and that the accidents are independent of each other. If it can be assumed that the process follows a Poisson distribution, then in a 350-day interval:

(a) What is the probability that there will be one accident?
(b) What is the probability that there are at most three days with an accident?

Chapter 2
Continuous Probability Distributions, Confidence Intervals, and Hypothesis Testing

In this chapter, we introduce continuous probability density functions: normal, lognormal, chi square, student t, f distribution, and Weibull. These probability density functions are then used to obtain the confidence intervals at a specified confidence level for the mean, differences in means, variance, ratio of variances, and difference in means for paired samples. These results are then extended to hypothesis testing where the p-value is introduced and the type I and type II errors are defined. The use of operating characteristic (OC) curves to determine the magnitude of these errors is illustrated. Also introduced is a procedure to obtain probability plots for the normal distribution as a visual means to confirm the normality assumption for data.

2.1 Introduction

Estimation is a procedure by which a sample from a population is used to investigate one or more properties of a population. A *point estimate* is the result of the estimation procedure that gives a single numerical value that is the most plausible value for the property of interest. For example, a reasonable point estimate of the mean of a population is the sample mean and for the variance the sample variance. However, the point estimate is itself a random variable since different sample sets will result in different numerical values of the point estimate. Therefore, it is also desirable to determine the variability associated with a point estimate. This variability is quantified with the generation of *confidence intervals*.

A confidence interval for a statistical quantity θ gives a range within which the parameter is expected to fall with a stated probability. This probability is called the *confidence level* and is the probability that the confidence interval contains the value

---

Supplementary Information The online version contains supplementary material available at [https://doi.org/10.1007/978-3-031-05010-7_2].

E. B. Magrab, *Engineering Statistics*, https://doi.org/10.1007/978-3-031-05010-7_2

Table 2.1 Probability distributions and their means and variances; all quantities are real and the special functions are in the table's footnote

Distribution	Probability density function	Cumulative density function	$E(x)$	Var(x)
Normal	$\frac{1}{\sqrt{2\pi}\sigma}\exp\left(-\frac{(x-\mu)^2}{2\sigma^2}\right)$	$\frac{1}{2}\mathrm{erfc}\left(\frac{\mu-x}{\sqrt{2}\sigma}\right)$	μ	σ^2
Lognormal ($x>0, \sigma>0$)	$\frac{1}{x\sigma\sqrt{2\pi}}\exp\left(-\frac{(\ln x-\mu)^2}{2\sigma^2}\right)$	$\frac{1}{2}\mathrm{erfc}\left(\frac{\ln x-\mu}{\sqrt{2}\sigma}\right)$	$e^{\mu+\sigma^2/2}$	$\left(e^{\sigma^2}-1\right)e^{2\mu+\sigma^2}$
Chi square χ^2 ($x>0, \nu=1, 2, \ldots$)	$\frac{2^{-\nu/2}}{\Gamma(\nu/2)}x^{\nu/2-1}e^{-x/2}$	$Q(\nu/2, x/2)$	ν	2ν
Student t ($\nu=1, 2, \ldots$)	$\frac{1}{\sqrt{\nu}B(\nu/2, 1/2)}\left(\frac{\nu}{\nu+x^2}\right)^{(\nu+1)/2}$	$\frac{1}{2}I_{\nu/(x^2+\nu)}(\nu/2, 1/2)$ $x\le 0$ $\frac{1}{2}(1+I_{x^2/(x^2+\nu)}(1/2, \nu/2))$ $x>0$	0 $\nu>1$	$\frac{\nu}{\nu-2}$ $\nu>2$
f ratio ($x>0$, $\nu_1, \nu_2=1, 2, \ldots$)	$\frac{\nu_2{}^{\nu_2/2}(\nu_1 x)^{\nu_1/2}(\nu_2+\nu_1 x)^{-(\nu_2+\nu_1)/2}}{xB(\nu_1/2, \nu_2/2)}$	$I_{\nu_1 x/(\nu_2+\nu_1 x)}(\nu_1/2, \nu_2/2)$	$\frac{\nu_2}{\nu_2-2}$ $\nu_2>2$	$\frac{2\nu_2^2(\nu_2+\nu_1-2)}{\nu_1(\nu_2-4)(\nu_2-2)^2}$ $\nu_2>4$
Weibull ($x\ge 0$, $\gamma>0, \beta>0$)	$\frac{\gamma}{\beta}\left(\frac{x}{\beta}\right)^{\gamma-1}\exp\left(-(x/\beta)^{\gamma}\right)$	$1-\exp\left(-(x/\beta)^{\gamma}\right)$	$\beta\Gamma\left(1+\frac{1}{\gamma}\right)$	$\beta^2\left(\Gamma\left(1+\frac{2}{\gamma}\right)-\Gamma\left(1+\frac{1}{\gamma}\right)^2\right)$

Rayleigh ($x \geq 0, \lambda > 0$)	$\frac{x}{\lambda^2} \exp\left(-\frac{x^2}{2\lambda^2}\right)$	$1 - \exp\left(-\frac{x^2}{2\lambda^2}\right)$	$\sqrt{\frac{\pi}{2}}\lambda$	$\left(2 - \frac{\pi}{2}\right)\lambda^2$
Exponential ($x \geq 0, \lambda > 0$)	$\lambda e^{-\lambda x}$	$1 - e^{-\lambda x}$	$\frac{1}{\lambda}$	$\frac{1}{\lambda^2}$

Special functions: Γ is the gamma function; Q is the regularized gamma function; B is the beta function; I is the regularized beta function; and erfc is the complementary error function. Their respective definitions are:

$$\Gamma(z) = \int_0^\infty t^{z-1} e^{-t} dt \quad \Gamma(a,z) = \int_z^\infty t^{a-1} e^{-t} dt \quad Q(a,z) = \frac{\Gamma(a,z)}{\Gamma(a)} \quad \operatorname{erfc}(z) = 1 - \frac{2}{\sqrt{\pi}} \int_0^z e^{-t^2} dt$$

$$B(a,b) = \int_0^1 t^{a-1}(1-t)^{b-1} dt \quad B_z(a,b) = \int_0^z t^{a-1}(1-t)^{b-1} dt \quad I_z(a,b) = \frac{B_z(a,b)}{B(a,b)}$$

of θ. The statistical quantity θ is usually a population mean, variance, ratio of variances, and the difference in means. The confidence interval's range is determined from an appropriate continuous probability distribution and a sample set from which $\bar{x}$ and s^2 are determined. It is unlikely that two sample sets from the same population will yield identical confidence intervals. However, if many sample sets are taken, a certain percentage of the resulting confidence intervals will contain θ. The percentage of these confidence intervals that contain θ is the confidence level of the interval. Confidence intervals will be illustrated as part of the discussions of the continuous probability distributions introduced in the following sections.

There are several probability density functions that are used to model and to analyze variability of sample means and variances. The ones that we shall consider are given in the first six rows of Table 2.1. It will be shown that several of the probability distributions provide the ability of estimating the population mean or variance with a confidence interval. These probability distributions and their corresponding confidence intervals also will be used in hypothesis testing discussed in Sect. 2.9 and a few of them in the analysis of variance technique discussed in Chap. 3 and 4.

The probability density functions and the cumulative density functions for the first six cases in Table 2.1 can be displayed with interactive graphic IG2–1.

2.2 Continuous Probability Distributions

2.2.1 *Introduction*

We introduce the continuous probability function by considering another way to look at the histogram given in Sect. 1.3.1. Recall from Eq. (1.2) that for discrete random variables the probability p_i of X being in the interval $\Delta_i = (y_{i+1} - y_i)$, $i = 1, 2, \ldots, m$, is

$$P\left(y_i \le X < y_{i+1}\right) = p_i = \frac{n_i}{n} \tag{2.1}$$

where y_j are the end points of the interval. We rewrite Eq. (2.1) as

$$P\left(y_i \le X < y_{i+1}\right) = \left(\frac{p_i}{\Delta_i}\right)\Delta_i$$

For a fixed and very large n, if we were to let Δ_i become smaller (m increases), we would expect the value of n_i to decrease and, therefore, p_i to decrease. However, what if the ratio p_i/Δ_i remains relatively unchanged such that in the limit

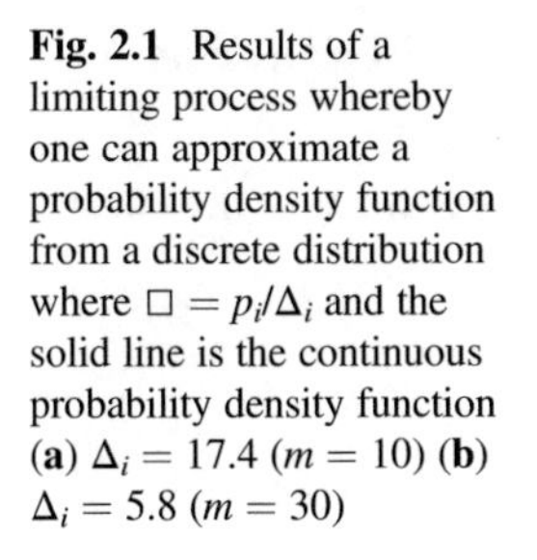

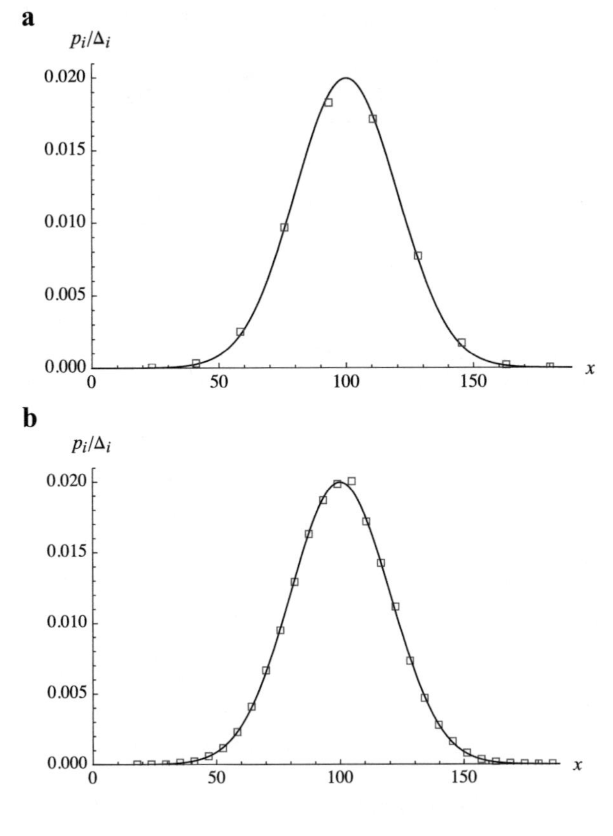

Fig. 2.1 Results of a limiting process whereby one can approximate a probability density function from a discrete distribution where □ = p_i/Δ_i and the solid line is the continuous probability density function (**a**) $\Delta_i = 17.4$ ($m = 10$) (**b**) $\Delta_i = 5.8$ ($m = 30$)

$$\lim_{\Delta_i \to 0} \left[\left(\frac{p_i}{\Delta_i} \right) \Delta_i \right] \overset{?}{\longrightarrow} f(x)dx \tag{2.2}$$

where $f(x)$ is the amount of probability in a very small interval dx, essentially a one-dimensional density. In this very small interval, the probability is $f(x)dx$. Notice that the probability at a specific value of X is zero since, in this case, $dx = 0$. Probability is a non-dimensional quantity. Then, if the units of x are denoted EU, then $f(x)$ has the units of EU^{-1} and, therefore, $f(x)dx$ is non-dimensional.

The idea behind Eq. (2.2) is shown in Fig. 2.1. In this case, we used a computer program to generate a very large number of random values ($n = 30{,}000$) having a specified distribution (the normal distribution, which is discussed in Sect. 2.3) with a mean $\mu = 100$ and a standard deviation $\sigma = 20$. The values range from $x_{\min} = 14.88$ to $x_{\max} = 188.91$ and two cases are considered: (a) $\Delta_i = 17.4$ ($m = 10$) and (b) $\Delta_i = 5.8$ ($m = 30$). It is seen that for both cases the values p_i/Δ_i are very close to the theoretical values given by the solid line.

Keeping this brief discussion in mind, we now define probability for continuous random variables.

2.2.2 Definitions Using Continuous Probability Distributions

For continuous random variables, the probability of an event occurring in the interval between a and b is defined as

$$P(a \leq X \leq b) = \int_a^b f(u)du \tag{2.3}$$

where $f(x)$ is called the *probability density function*. Thus, the probability $P(a \leq X \leq b)$ is the area under the curve described by $f(x)$ in this interval. In addition, the probability density function has the property that for all real X, $f(x) \geq 0$, and

$$\int_{-\infty}^{\infty} f(u)du = 1 \tag{2.4}$$

When $a \to -\infty$ in Eq. (2.3), we have the definition of the *cumulative distribution function*

$$F(x) = P(X \leq x) = \int_{-\infty}^{x} f(u)du \qquad -\infty \leq x \leq \infty \tag{2.5}$$

From calculus, it is noted that

$$\frac{dF(x)}{dx} = f(x) \tag{2.6}$$

Expected value The expected value for a continuous distribution is defined as (recall Eq. (1.10))

$$E(X) = \mu = \int_{-\infty}^{\infty} uf(u)du \tag{2.7}$$

which is the mean of X.

Variance The variance is defined as (recall Eq. (1.15))

$$\mathrm{Var}(X) = E\left((X-\mu)^2\right) = \int_{-\infty}^{\infty} (u-\mu)^2 f(u)du = \int_{-\infty}^{\infty} u^2 f(u)du - \mu^2 \tag{2.8}$$

Median The median is obtained by determining the value of m that satisfies

$$\int_{-\infty}^{m} f(u)du = 0.5 \tag{2.9}$$

It is seen from these equations that once $f(x)$ is known, only integration is required to determine these various quantities. The probability density functions that we shall consider in this chapter are given in Table 2.1 along with their respective cumulative distribution function, expected value, and variance.

2.3 Normal Distribution

2.3.1 Normal Distribution

From Table 2.1, the normal probability density function is

$$f_N(x) = N\left(x;\mu,\sigma^2\right) = \frac{1}{\sqrt{2\pi}\sigma}\exp\left(-\frac{(x-\mu)^2}{2\sigma^2}\right) \tag{2.10}$$

where μ and σ^2 are constants. Eq. (2.10) is a symmetric function about $x = \mu$. We introduce the following notation to define the following probability

$$P(X \ge x) = \frac{1}{\sqrt{2\pi}\sigma}\int_{x}^{\infty} \exp\left(-\frac{(y-\mu)^2}{2\sigma^2}\right)dy = \frac{1}{\sqrt{2\pi}}\int_{(x-\mu)/\sigma}^{\infty} e^{-u^2/2}du \tag{2.11}$$

Setting

$$z = \frac{x-\mu}{\sigma} \quad \text{and} \quad Z = \frac{X-\mu}{\sigma} \tag{2.12}$$

which are called the *standard normal random variables*, Eq. (2.11) can be written as

$$P(X \geq x) = P\left(\frac{X-\mu}{\sigma} \geq \frac{x-\mu}{\sigma}\right) = \frac{1}{\sqrt{2\pi}} \int_{(x-\mu)/\sigma}^{\infty} e^{-u^2/2} du$$

or

$$\Psi_N(z) = P(Z \geq z) = \int_{z}^{\infty} f_N(u) du \tag{2.13}$$

where

$$f_N(u) = \frac{e^{-u^2/2}}{\sqrt{2\pi}} \tag{2.14}$$

is called the *standard normal distribution*. Eq. (2.14) is symmetric about $\mu = 0$ and, therefore, the mean equals the median.

Consider the case where $\Psi_N(z) = \alpha$. There are two interpretations to this equation. In the first instance, one selects z and the evaluation of the integral results in α. This interpretation will be used in Sect. 2.9.2 to determine a probability called the p-value. In the second instance, one specifies α and then determines that value of $z = z_\alpha$ such that the integral evaluates to α. In this case, we are solving the inverse problem

$$\Psi_N^{-1}(\alpha) = z_\alpha \tag{2.15}$$

where the superscript '–1' indicates that the inverse operation has been performed. For this case, z_α is called the *critical value*. Then, we can write Eq. (2.13) as

$$\Psi_N(z_\alpha) = \alpha = P(Z \geq z_\alpha) = \int_{z_\alpha}^{\infty} f_N(u) du \tag{2.16}$$

We note that

$$\int_{-\infty}^{\infty} f_N(u) du = 1$$

Then, Eq. (2.16) can be written as

$$P(Z \leq z_\alpha) = 1 - P(Z \geq z_\alpha) = 1 - \alpha \tag{2.17}$$

where $P(Z \leq z_\alpha)$ is the *cumulative distribution function* for the normal distribution and is given by

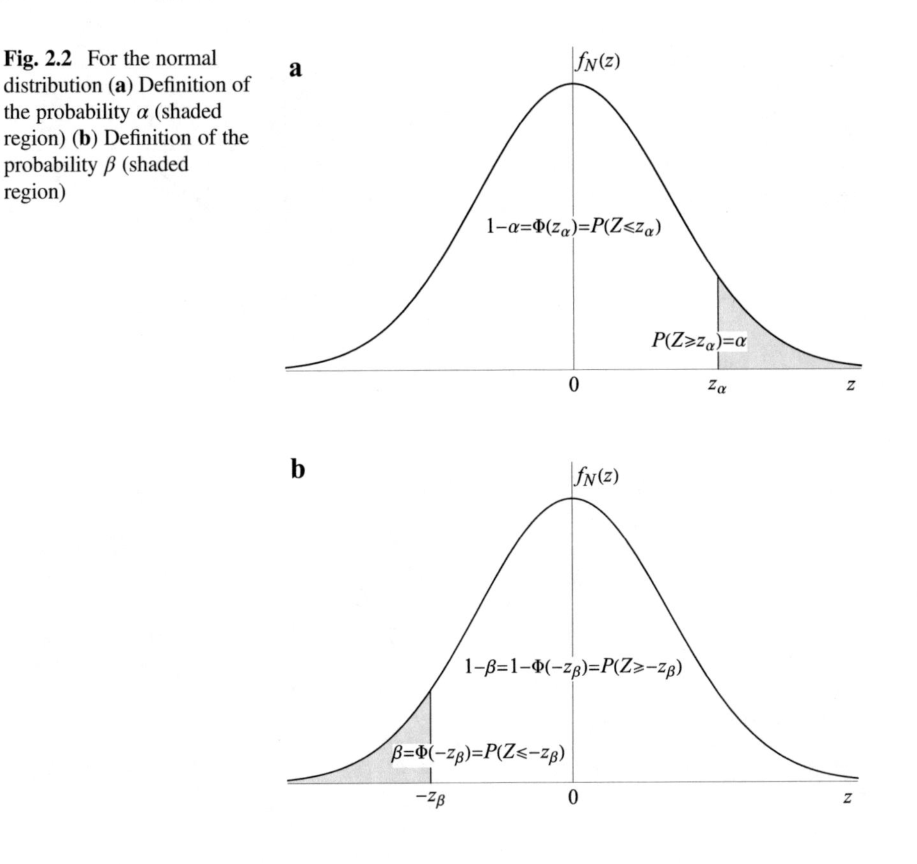

Fig. 2.2 For the normal distribution (**a**) Definition of the probability α (shaded region) (**b**) Definition of the probability β (shaded region)

$$\Phi(z) = P(Z \le z) = \frac{1}{\sqrt{2\pi}} \int_{-\infty}^{z} e^{-u^2/2} du \tag{2.18}$$

The quantity $\Phi(z)$ is a tabulated function. Then using Eq. (2.18) in Eq. (2.17), we obtain

$$P(Z \ge z_\alpha) = 1 - P(Z \le z_\alpha) = 1 - \Phi(z_\alpha) = \alpha \tag{2.19}$$

These definitions are shown in Fig. 2.2.

Like Eq. (2.15), Eq. (2.18) can also be used in an inverse manner. This inverse operation for two cases is denoted

$$\begin{aligned} z_\alpha &= \Phi^{-1}(1-\alpha) \\ -z_\beta &= \Phi^{-1}(\beta) \end{aligned} \tag{2.20}$$

where $-z_\beta$ is shown in Fig. 2.2b. The critical values of z_α that satisfy Eq. (2.20) are shown in Fig. 2.3 for the stated values of α.

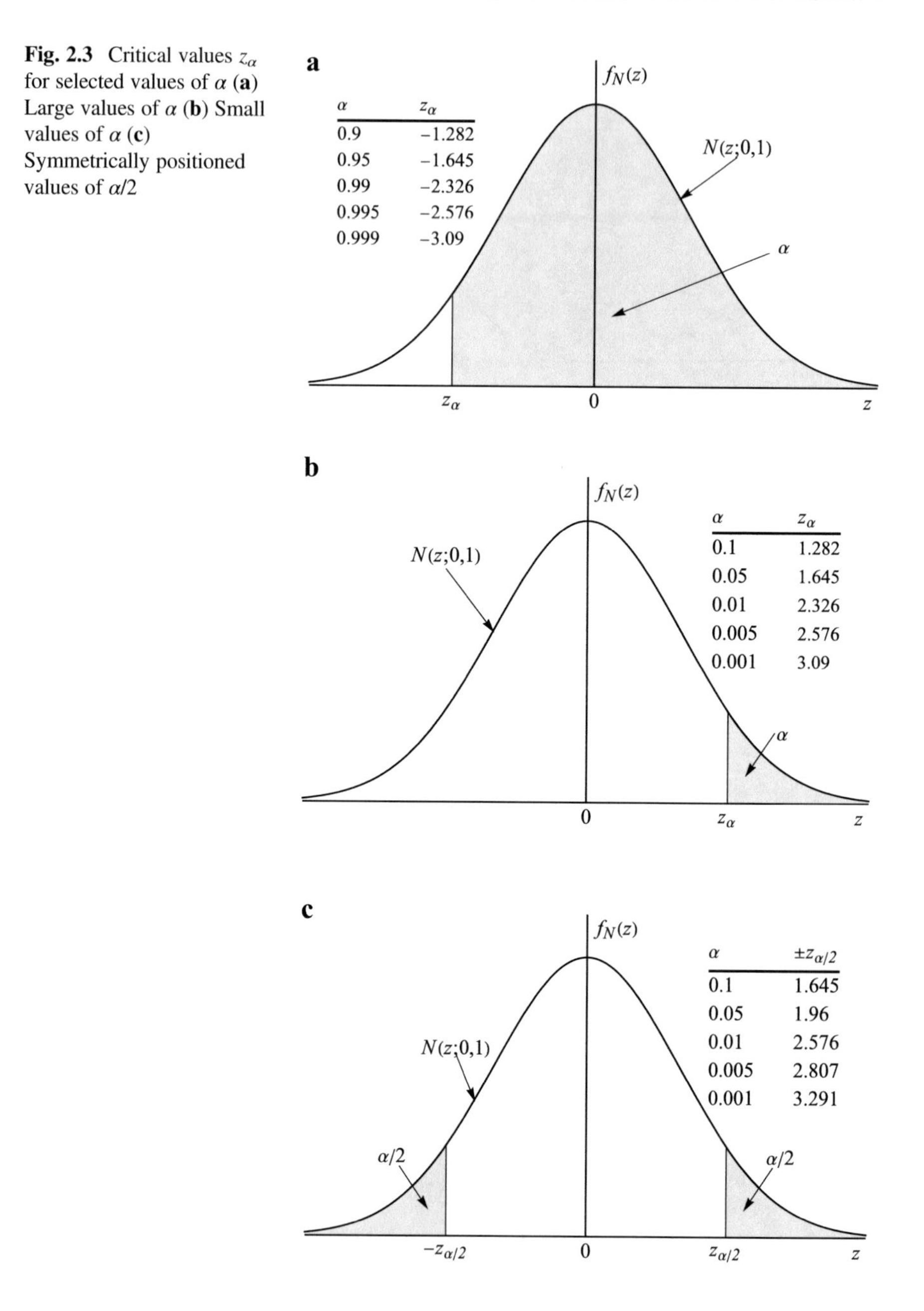

Fig. 2.3 Critical values z_α for selected values of α (**a**) Large values of α (**b**) Small values of α (**c**) Symmetrically positioned values of $\alpha/2$

The evaluation of $\Psi_N(z_\alpha) = \alpha$, $\Psi_N^{-1}(\alpha) = z_\alpha$, which gives the critical value z_α, and $\Phi(z)$, which gives the value for the cumulative distribution function, can be obtained with interactive graphic IG2–2.

The expected value of $f(x)$ is obtained from the evaluation of Eq. (2.7), which gives

$$E(X) = \frac{1}{\sqrt{2\pi}\sigma} \int_{-\infty}^{\infty} u \exp\left(-(u-\mu)^2/(2\sigma^2)\right) du = \mu \tag{2.21}$$

and the variance is obtained from the evaluation of Eq. (2.8) as

$$\text{Var}(X) = \frac{1}{\sqrt{2\pi}\sigma} \int_{-\infty}^{\infty} u^2 \exp\left(-(u-\mu)^2/(2\sigma^2)\right) du - \mu^2 = \sigma^2 \tag{2.22}$$

In obtaining Eq. (2.22), we have used Eq. (2.21). Thus, the parameters μ and σ^2 appearing in Eq. (2.10) represent, respectively, the mean and variance of this distribution. In Sect. A.3 of Appendix A, it is shown that for the normal distribution the mean and variance are independent parameters. Therefore, the normal probability density function is defined by two independent constants: the mean and the variance.

Using Eqs. (2.12) and (2.14), the expected value of Z is

$$E(Z) = \mu = \frac{1}{\sqrt{2\pi}} \int_{-\infty}^{\infty} u \exp\left(-u^2/2\right) du = 0 \tag{2.23}$$

and its variance is

$$\text{Var}(Z) = \frac{1}{\sqrt{2\pi}} \int_{-\infty}^{\infty} u^2 \exp\left(-u^2/2\right) du = 1 \tag{2.24}$$

Therefore, Eq. (2.14) can be written as.

$$f_N(z) = N(z; 0, 1) = \frac{e^{-z^2/2}}{\sqrt{2\pi}} \tag{2.25}$$

The implications of the transformation given by Eq. (2.12) are shown in Fig. 2.4. In Fig. 2.4a, we have given three different probability density functions in terms of their respective sample values x, mean μ, and variance σ^2. In Fig. 2.4b, we have given $N(z;0,1)$, which is the same for all three functions given in Fig. 2.3a. It is mentioned again that each probability density function in Fig. 2.3a maps into the single probability density function given by $N(z;0,1)$ in Fig. 2.3b. However, the area under the curve determined by either $N(x;\mu,\sigma^2)$ or $N(z;0,1)$, is the same; that is, using

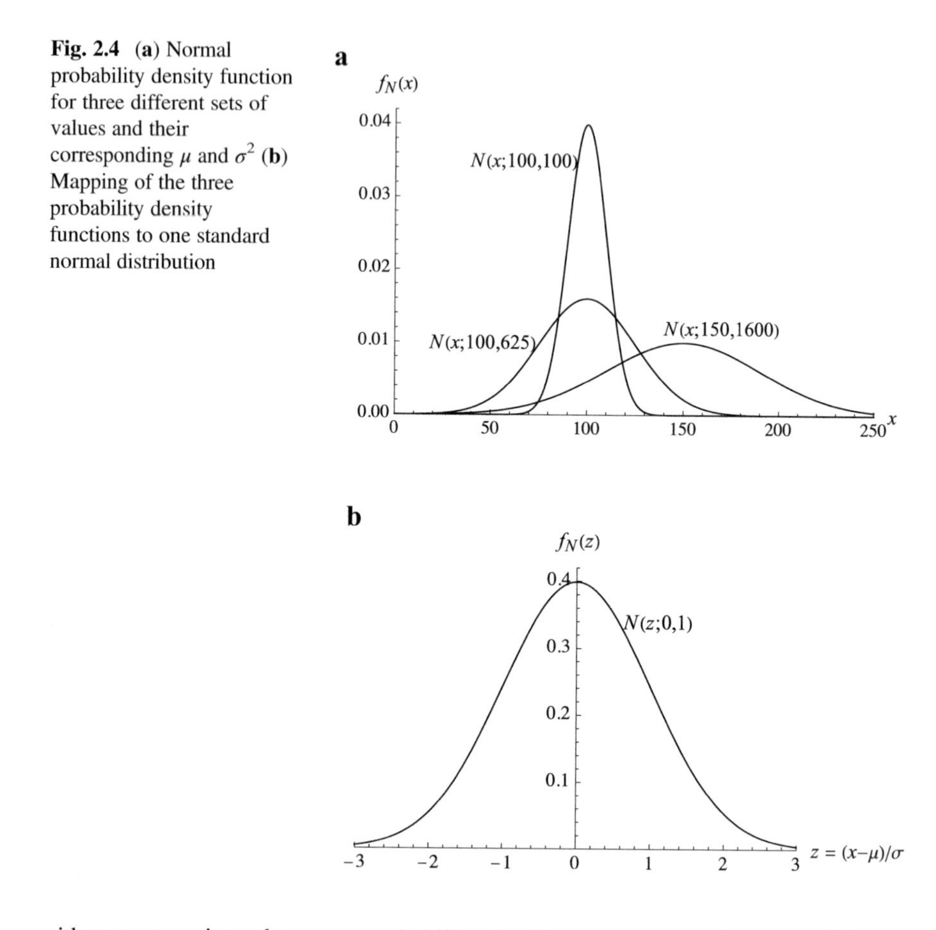

Fig. 2.4 (**a**) Normal probability density function for three different sets of values and their corresponding μ and σ^2 (**b**) Mapping of the three probability density functions to one standard normal distribution

either x or z gives the same probability. In this regard, note the very different magnitudes of their respective vertical and horizontal scales in these figures.

For $z_\alpha > z_\beta$, consider the probability statement

$$\begin{aligned} P\left(z_\beta \le Z \le z_\alpha\right) &= P(Z \le z_\alpha) - P\left(Z \le z_\beta\right) = \Phi(z_\alpha) - \Phi\left(z_\beta\right) \\ &= 1 - \alpha - \beta \end{aligned} \tag{2.26}$$

which is illustrated in Fig. 2.5 for $z_\beta < 0$. In the case where $z_\beta < 0$ and $\alpha = \beta$, we see that because $N(z;0,1)$ is symmetrical about $z = 0$, $\Phi(-z_\alpha) = 1 - \Phi(z_\alpha)$. Then, Eq. (2.26) becomes

$$P(-z_\alpha \le Z \le z_\alpha) = 2\Phi(z_\alpha) - 1 = 1 - 2\alpha \tag{2.27}$$

Equation (2.27) can be written as

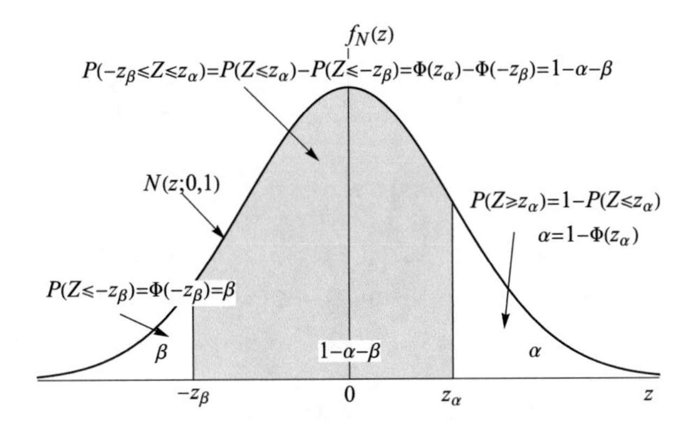

Fig. 2.5 Probabilities obtained from the standard normal probability distribution for a given α and β and their corresponding critical values

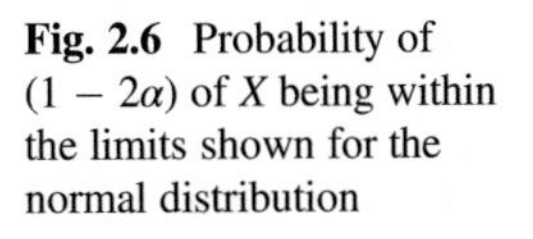

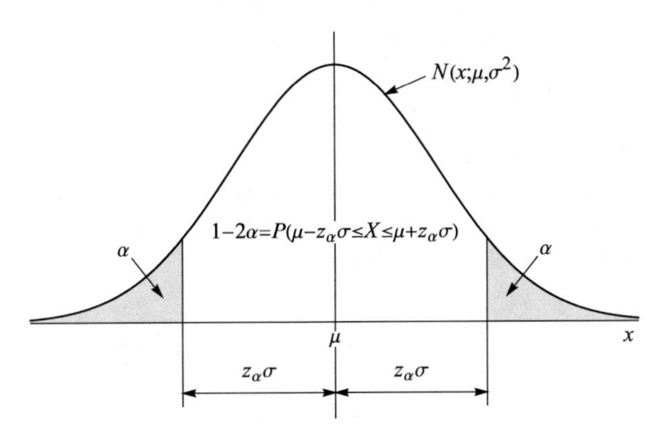

Fig. 2.6 Probability of $(1 - 2\alpha)$ of X being within the limits shown for the normal distribution

$$P\left(-z_\alpha \leq \frac{X-\mu}{\sigma} \leq z_\alpha\right) = 1 - 2\alpha$$

or

$$P(\mu - z_\alpha\sigma \leq X \leq \mu + z_\alpha\sigma) = 1 - 2\alpha \tag{2.28}$$

which is shown in Fig. 2.6.

From Eq. (2.12), we see that $x = z\sigma + \mu$; therefore, $z = \pm 1$ corresponds to x being one standard deviation from the mean, $z = \pm 2$ corresponds to being two standard deviations from the mean, and so on. Then the probability $P(-k \leq Z \leq k)$, $k = 1, 2, 3$ of Z being in these intervals is obtained from Eqs. (2.26) and (2.18) as

$$P(-k \le Z \le k) = \Phi(k) - \Phi(-k) = \frac{1}{\sqrt{2\pi}} \int_{-\infty}^{k} e^{-u^2/2} du - \frac{1}{\sqrt{2\pi}} \int_{-\infty}^{-k} e^{-u^2/2} du$$

For $k = 1, 2, 3$, we find that

$$\begin{aligned} P(-1 \le Z \le 1) &= \Phi(1) - \Phi(-1) = 0.8413 - 0.15866 = 0.6827 \\ P(-2 \le Z \le 2) &= \Phi(2) - \Phi(-2) = 0.9773 - 0.02275 = 0.9545 \\ P(-3 \le Z \le 3) &= \Phi(3) - \Phi(-3) = 0.9987 - 0.00135 = 0.9973 \end{aligned} \tag{2.29}$$

where the values for $\Phi(\pm k)$ were obtained from interactive graphic IG2–2. Thus, about 68% of the standard normal variable lies within ± 1 standard deviation, 95% within ± 2 standard deviations, and 99.7% within ± 3 standard deviations.

We now illustrate these results with the following example.

Example 2.1

Given that a population has a normal distribution with a mean $\mu = 100$ and a standard deviation $\sigma = 16$.

(a) We shall determine the probability that a sample will have a value in the interval between $x_L = 84$ and $x_U = 132$. Using Eq. (2.12), we convert these values to their standard normal form and obtain

$$\begin{aligned} z_U &= \frac{x_U - \mu}{\sigma} = \frac{132 - 100}{16} = 2 \\ z_L &= \frac{x_L - \mu}{\sigma} = \frac{84 - 100}{16} = -1 \end{aligned}$$

Then, from Eq. (2.26), we have

$$\begin{aligned} P(z_L \le Z \le z_U) &= \Phi(z_U) - \Phi(z_L) = \Phi(2) - \Phi(-1) \\ &= 0.9773 - 0.1587 = 0.8186 \end{aligned}$$

where Φ was obtained from interactive graphic IG2–2.

(b) We shall determine the critical value of x for which the probability of exceeding that value is 0.15. Then, from interactive graphic IG2–2, we find that $\Phi_N^{-1}(0.15) = z_{0.15} = 1.0364$. Therefore, from Eq. (2.12)

$$x_{0.15} = \sigma z_{0.15} + \mu = 16 \times 1.0364 + 100 = 116.58$$

2.3.2 Combining Independent Normally Distributed Random Variables

We now consider the sum of n independent random variables X_i each of which is normally distributed; that is, $X_i \sim N(\mu_i, \sigma_i^2)$, $i = 1, 2, \ldots, n$, where '~' means 'distributed as', $\mu_i = E(X_i)$, $\sigma_i^2 = \text{var}(X_i)$ and, for convenience, we have shortened $N(X_i; \mu_i, \sigma_i^2)$ to $N(\mu_i, \sigma_i^2)$. Then the following linear combination is also normally distributed

$$Y = \sum_{i=1}^{n} a_i X_i + b \sim N(\mu, \sigma^2) \tag{2.30}$$

where, from Eqs. (1.30) and (1.31),

$$\begin{aligned} \mu &= \sum_{i=1}^{n} a_i \mu_i + b \\ \sigma^2 &= \sum_{i=1}^{n} a_i^2 \sigma_i^2 \end{aligned} \tag{2.31}$$

Consider the case when that $a_i = 1/n$ and $b = 0$. Then, Eq. (2.30) becomes

$$Y = \frac{1}{n} \sum_{i=1}^{n} X_i = \overline{X} \tag{2.32}$$

If $\mu = \mu_i = E(X_i)$ and $\sigma^2 = \sigma_i^2 = \text{var}(X_i)$, $i = 1, 2, \ldots, n$; that is, they are independently and identically distributed as $X_i \sim N(\mu, \sigma^2)$, then the expected value and variance of Y, respectively, are obtained using Eqs. (2.32) as

$$\begin{aligned} E(Y) = E(\overline{X}) &= \frac{1}{n} \sum_{i=1}^{n} E(X_i) = \frac{1}{n} \sum_{i=1}^{n} \mu = \mu \\ \text{Var}(Y) = \text{Var}(\overline{X}) &= \frac{1}{n^2} \sum_{i=1}^{n} \sigma_i^2 = \frac{1}{n^2} \sum_{i=1}^{n} \sigma^2 = \frac{\sigma^2}{n} \end{aligned} \tag{2.33}$$

Therefore, the mean is normally distributed as $\overline{X} \sim N(\mu, \sigma^2/n)$. In this case, one writes the standard variable as

$$Z_{\bar{x}} = \frac{\bar{x} - \mu}{\sigma/\sqrt{n}} \tag{2.34}$$

Using Eq. (2.27) with $\alpha = \alpha/2$, the probability statement for the mean becomes

$$P\left(-z_{\alpha/2} \le Z_{\bar{x}} \le z_{\alpha/2}\right) = 1 - \alpha$$

or using Eq. (2.34)

$$P\left(\bar{x} - z_{\alpha/2}\sigma/\sqrt{n} \le \mu \le \bar{x} + z_{\alpha/2}\sigma/\sqrt{n}\right) = 1 - \alpha \tag{2.35}$$

Equation (2.35) states that the population mean μ has a $(1 - \alpha)$ probability of being within the limits shown. Thus, we define the $100(1 - \alpha)\%$ *confidence interval* for the population mean as

$$\bar{x} - z_{\alpha/2}\sigma/\sqrt{n} \le \mu \le \bar{x} + z_{\alpha/2}\sigma/\sqrt{n} \tag{2.36}$$

when σ is known. It is noted that the end points of the interval in Eqs. (2.36) are random variables since $\bar{x}$ is random variable. Therefore, the interpretation of this expression is that there is a probability $\alpha/2$ that when an average is computed from n samples taken from the population $\mu < \bar{x} - z_{\alpha/2}\sigma/\sqrt{n}$ or a probability $\alpha/2$ that $\mu > \bar{x} + z_{\alpha/2}\sigma/\sqrt{n}$. In other words, if we were to repeat this test with n samples a sufficcient numbr of times, $100(1 - \alpha)\%$ of the time the interval will contain μ.

The one-sided $100(1 - \alpha)\%$ upper confidence limit is obtained by replacing $z_{\alpha/2}$ with z_α in Eq. (2.36), which yields

$$\mu \le \bar{x} + z_\alpha\sigma/\sqrt{n} \tag{2.37}$$

and the lower confidence limit is

$$\bar{x} - z_\alpha\sigma/\sqrt{n} \le \mu \tag{2.38}$$

These results are summarized as Case 1 in Table 2.2. For specific numerical values for $\bar{x}$, σ, and n, Eqs. (2.36) to (2.38) can be determined using interactive graphic IG2–3.

Note from Eq. (2.36) that as α becomes smaller; that is, our confidence level $(1 - \alpha)100\%$ becomes higher, our confidence interval becomes larger. This is because as α becomes smaller the corresponding critical value z_α becomes larger.

We see from Eqs. (2.36) to (2.38) that there are two ways to decrease the confidence intervals: increase α or increase n. Since we often want a confidence level that is as high as practical, increasing α is not a good alternative. Therefore, the only option is to increase the number of samples n. The confidence interval for the two-sided confidence interval is governed by the quantity

$$\Delta_0 = z_{\alpha/2}\frac{\sigma}{\sqrt{n}} \tag{2.39}$$

In other words, the variation about the population mean is $\pm\Delta_0$. If Δ_0 is given a value, then for a given α we choose

Table 2.2 Summary of confidence intervals on the mean and variance presented in Sect. 2.3.2 to 2.6[a]

			100(1 − α)% confidence interval		
			$\widehat{\theta}-q \le \theta \le \widehat{\theta}+q$	$\theta \le \widehat{\theta}+q'$ or $\widehat{\theta}-q' \le \theta$	
Case	θ	$\widehat{\theta}$	q	q'	Assumptions
1	μ	$\bar{x}$	$z_{\alpha/2}\sigma/\sqrt{n}$	$z_{\alpha}\sigma/\sqrt{n}$	σ^2 known
2	μ	$\bar{x}$	$t_{\alpha/2,\nu}s/\sqrt{n}$	$t_{\alpha,\nu}s/\sqrt{n}$	σ^2 unknown
3	$\mu_1-\mu_2$	$\bar{x}_1-\bar{x}_2$	$z_{\alpha/2}\sqrt{\sigma_1^2/n_1+\sigma_2^2/n_2}$	$z_{\alpha}\sqrt{\sigma_1^2/n_1+\sigma_2^2/n_2}$	σ_1^2, σ_2^2 known
4	$\mu_1-\mu_2$	$\bar{x}_1-\bar{x}_2$	$t_{\alpha/2,\nu}S_p\sqrt{1/n_1+1/n_2}$ $\nu=n_1+n_2-2$ S_p from Eq.(2.92)	$t_{\alpha,\nu}S_p\sqrt{1/n_1+1/n_2}$ $\nu=n_1+n_2-2$ S_p from Eq.(2.92)	$\sigma_1^2=\sigma_2^2=\sigma^2$ unknown
5	$\mu_1-\mu_2$	$\bar{x}_1-\bar{x}_2$	$t_{\alpha/2,\nu}\sqrt{s_1^2/n_1+s_2^2/n_2}$ ν from Eq.(2.96)	$t_{\alpha,\nu}\sqrt{s_1^2/n_1+s_2^2/n_2}$ ν from Eq.(2.96)	$\sigma_1^2 \ne \sigma_2^2$ unknown
			100(1−α)% confidence interval		
			$q_1\widehat{\theta} \le \theta \le q_2\widehat{\theta}$		
			q_1	q_2	
6	σ^2	s^2	$\dfrac{(n-1)}{\chi^2_{\alpha/2,n-1}}$	$\dfrac{(n-1)}{\chi^2_{1-\alpha/2,n-1}}$	
7	$\dfrac{\sigma_1^2}{\sigma_2^2}$	$\dfrac{s_1^2}{s_2^2}$	$f_{1-\alpha/2,\nu_2,\nu_1}$	$f_{\alpha/2,\nu_2,\nu_1}$	

[a]The confidence intervals in this table can be numerically evaluated with interactive graphic IG2–3

$$n = \left\lceil \left(z_{\alpha/2} \frac{\sigma}{\Delta_0} \right)^2 \right\rceil \tag{2.40}$$

where $\lceil x \rceil$ is the ceiling function indicating that x will become the smallest integer greater than x.

We now illustrate these results with the following examples.

Example 2.2
Consider two independent processes, each of which is producing one of two mating parts. The mating parts are required to maintain a gap that is to stay within a specified range that is greater than zero. One part has a mean $\mu_1 = 25.00$ mm and a standard deviation $\sigma_1 = 0.06$ mm. For the other part, $\mu_2 = 25.20$ mm and $\sigma_2 = 0.04$ mm. If it is assumed that the measured values from both processes are normally distributed, then we use Eq. (2.31) with $a_1 = 1$, $a_2 = -1$, and $b = 0$, to determine the average gap between the parts and its variance. Thus,

$$\mu = 25.20 - 25.00 = 0.20$$
$$\sigma^2 = (0.06)^2 + (0.04)^2 = 0.0052$$

and the linear combination $Y \sim N(0.20, 0.0052)$. We shall determine the probability that the gap is between $y_L = 0.15$ mm and $y_U = 0.27$ mm. Using Eq. (2.12), we find that

$$z_L = \frac{y_L - \mu}{\sigma} = \frac{0.15 - 0.20}{0.07211} = -0.6933$$
$$z_U = \frac{y_U - \mu}{\sigma} = \frac{0.27 - 0.20}{0.07211} = 0.971$$

Then, from Eq. (2.26),

$$P(z_L \le Y \le z_U) = \Phi(z_U) - \Phi(z_L) = \Phi(0.971) - \Phi(-0.6933)$$
$$= 0.8342 - 0.2441 = 0.590$$

where Φ was obtained from interactive graphic IG2–2.

The probability that the gap is greater than zero is, from Eq. (2.19),

$$P(Y \ge 0) = 1 - \Phi\left(\frac{0 - 0.20}{0.07221}\right) = 1 - \Phi(-2.774) = 1 - 0.00278 = 0.9972$$

where Φ was obtained from interactive graphic IG2–2.

Example 2.3
An experiment obtains a mean value of 2.5 for 30 observations. If the population standard deviation is 0.31, then we shall determine the confidence interval of the mean at the 95% level. Using Eq. (2.36), we obtain

$$2.5 - \frac{1.96 \times 0.31}{\sqrt{30}} \le \mu \le 2.5 + \frac{1.96 \times 0.31}{\sqrt{30}}$$
$$2.389 \le \mu \le 2.611$$

where from Fig. 2.3c we find that $z_{0.025} = 1.96$.

This result can be verified with interactive graphic IG2–3.

It is seen from this result that $\Delta_0 = (2.611–2.389)/2 = 0.111$. If we require that $\Delta_0 = 0.08$, then from Eq. (2.40) the number of observations must be increased to

$$n = \left\lceil \left(1.96\frac{0.31}{0.08}\right)^2 \right\rceil = \lceil 57.68 \rceil = 58$$

As a check, we use Eq. (2.39) and find that $\Delta_0 = 1.96 \times 0.31/\sqrt{58} = 0.07978 < 0.08$.

2.3.3 *Probability Plots for the Normal Distribution*

We now introduce a method that can be used to obtain a qualitative visual means of determining whether a set of data is normally distributed. As will be seen subsequently, it is often necessary that the data being analyzed are normally distributed, for the relations used in the analysis require normality in order that they be applicable.

The method constructs a probability graph in the following manner. Consider the data set $S = \{x_1, x_2, x_3, \ldots, x_n\}$. The data are ordered from smallest to largest value so that S becomes $\widetilde{S} = \{\widetilde{x}_1, \widetilde{x}_2, \widetilde{x}_3, \ldots, \widetilde{x}_n\}$, where $\widetilde{x}_{i+1} \ge \widetilde{x}_i \; i = 1, 2, \ldots, n-1$. Recall Sect. 1.3.1. Associated with each $\widetilde{x}_i$ is the approximate cumulative probability $p_i = P(X \le \widetilde{x}_i)$ that is given by one of several relations: either

$$p_i = \frac{i - 0.5}{n} \quad i = 1, 2, \ldots, n \tag{2.41}$$

or

$$p_i = \begin{cases} 1 - 0.5^{1/n} & i = 1 \\ (i - 0.3175)/(n + 0.365) & i = 2, 3, \ldots, n-1 \\ 0.5^{1/n} & i = n \end{cases} \tag{2.42}$$

or

$$p_i = \frac{i - 0.375}{n + 0.25} \quad i = 1, 2, \ldots, n \tag{2.43}$$

or

$$p_i = \frac{i}{n+1} \quad i = 1, 2, \ldots, n \tag{2.44}$$

The $\tilde{x}_i$ are plotted on the x-axis and the corresponding p_j on the y-axis.[1] To make an interpretation of these data values easier, often a straight line is fit to them and included in the plot. In theory, if the data are normally distributed, the data points would all lie on this line. In practice, they rarely do, but if the data are indeed normally distributed, most of the values will be in close proximity to the line. When a number of values diviate significantly from the straight line, the data may not be normally distributed.

Another way that the data are presented is in terms of the normal standard variable. In this case, the y-axis values become $y_i = \Phi^{-1}(p_i)$ and the x-axis values become $z_i = (\tilde{x}_i - \overline{x})/s$ so that both axes ranges are roughly within ± 3. The straight line is fit to these data as discussed in Sect. 3.2. When a straight line is fit to these transformed data, if the slope of the line is close to one and if most of the plotted values are in close proximity to this line then the data can be considered normally distributed. See also the discussion regarding Case 4 of Table 3.1 and Fig. 3.2. These probability plots will be used in regression analysis and in the analysis of variance, which are discussed in Chap. 3.

To illustrate the method, we have chosen to plot $\Phi^{-1}(p_i)$ versus z_i, which is shown in Fig. 2.7 with p_i given by Eq. (2.41).[2] The data consist of 30 samples with a mean of 100.59 and a standard deviation of 5.038. The minimum value of the data is 90.93 and its maximum value is 109.56. In Fig. 2.7a, the slope of the fitted line is 0.985; thus, from the way in which the data are close to the line and that the slope of the line is almost one, these data can be considered normally distributed. The data in Fig. 2.7b are an example of data that are not normally distributed. Experience has found that when the sample size is less than 30, only very large deviations from linearity should be considered an indication of nonnormality.

2.3.4 *Central Limit Theorem*

One of the most useful theorems in statistics is the *central limit theorem*, which provides a way of approximating the probability of the average of a sampled population whose probability distribution is unknown. If we take a random sample $S = \{X_1, X_2, \ldots, X_n\}$ from a population with an unknown distribution and for which each sample has the identical distribution with mean μ and variance σ^2, then as n becomes large the distribution of their average value $\overline{X}$ is always approximately

---

[1] Sometimes these values are displayed with p_i on the x-axis and $\tilde{x}_i$ on the y-axis.

[2] For these data, there is very little difference visually when comparing the results using Eq. (2.41) to those using Eqs. (2.42) to (2.44).

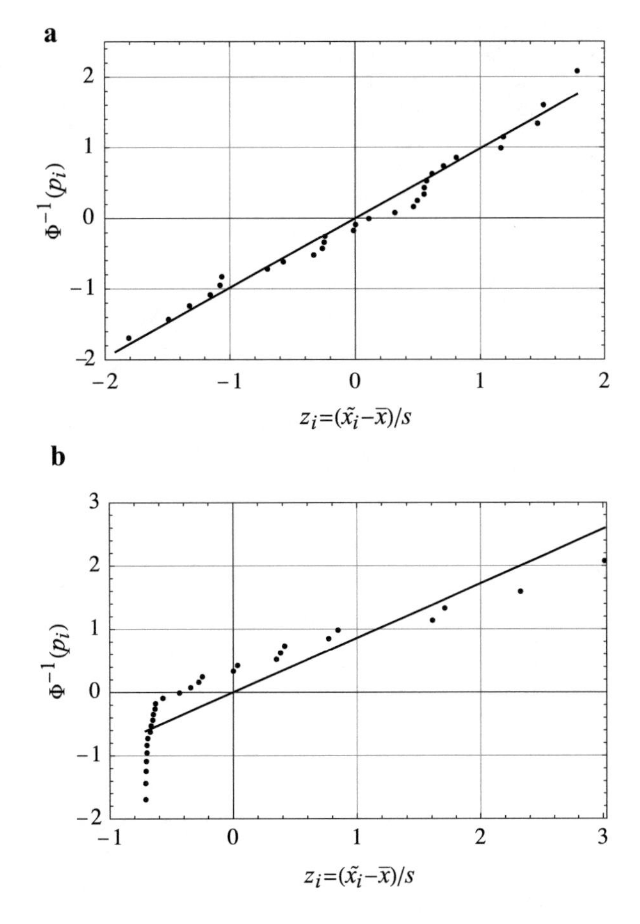

Fig. 2.7 Probability plots (**a**) Data that are normally distributed (**b**) Data that are not normally distributed

distributed as $N(\mu, \sigma^2/n)$. In terms of the standard normal random variable $\overline{Z}_{\bar{x}}$ given by Eq. (2.34), we have that as n becomes very large $\overline{X} \sim N(0, 1)$.

In practice, this approximation will generally be applicable for $n > 30$ provided that the distribution is not too skewed. The more skewed the distribution, the more n must exceed 30. If $n < 30$, then this approximation will work if the distribution is not too different from the normal distribution.

Thus, when the sample sizes are large as indicated above, one may be able to use Eqs. (2.36) to (2.38).

2.3.5 Lognormal Distribution

A random variable X is said to have a *lognormal* distribution if the distribution $Y = \ln(X)$ (or, equivalently, $X = e^Y$) has a normal distribution $Y \sim N(\mu,\sigma^2)$. The lognormal probability density function is given by

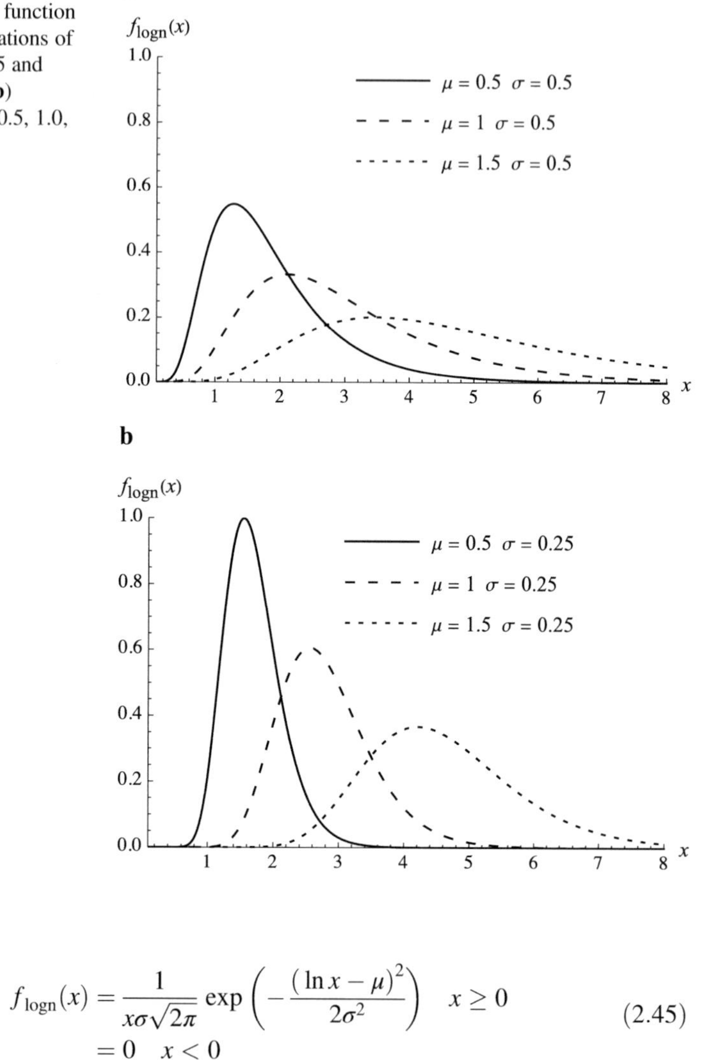

Fig. 2.8 The lognormal probability density function for several combinations of μ and σ (**a**) $\sigma = 0.5$ and $\mu = 0.5, 1.0, 1.5$ (**b**) $\sigma = 0.25$ and $\mu = 0.5, 1.0, 1.5$

$$f_{\text{logn}}(x) = \frac{1}{x\sigma\sqrt{2\pi}} \exp\left(-\frac{(\ln x - \mu)^2}{2\sigma^2}\right) \quad x \geq 0$$
$$= 0 \quad x < 0 \tag{2.45}$$

Equation (2.45) is plotted in Fig. 2.8 for several combinations of μ and σ, where μ and σ^2 are the mean and variance of Y not X. Populations that are candidates for a lognormal transformation often can be determined from their histogram. At a minimum, the histograms of X will only contain values that are greater than zero and there will be seemingly many outliers in the positive direction.

The cumulative distribution function can be written as

$$P(X \le x) = \frac{1}{\sigma\sqrt{2\pi}} \int_0^x y^{-1} \exp\left(-\frac{(\ln y - \mu)^2}{2\sigma^2}\right) dy$$
$$= \frac{1}{\sqrt{2\pi}} \int_{-\infty}^{(\ln x - \mu)/\sigma} e^{-u^2/2} du \tag{2.46}$$
$$= \Phi((\ln x - \mu)/\sigma)$$

Then, if we set

$$z = \frac{\ln x - \mu}{\sigma} \qquad Z = \frac{\ln X - \mu}{\sigma} = \frac{Y - \mu}{\sigma} \tag{2.47}$$

Eq. (2.46) can be written as

$$P(X \le x) = P(e^Y \le x) = P(Y \le \ln x)$$
$$= P\left(\frac{Y - \mu}{\sigma} \le \frac{\ln x - \mu}{\sigma}\right) = P(Z \le z) \tag{2.48}$$
$$= \Phi(z) = \Phi\left(\frac{\ln x - \mu}{\sigma}\right)$$

Therefore, Eq. (2.20) applies so that for a specified value for α the value for which $X \le x_\alpha$ is

$$\frac{\ln x_\alpha - \mu}{\sigma} = \Phi^{-1}(1 - \alpha) \to x_\alpha = e^{\mu + \sigma\Phi^{-1}(1-\alpha)} \tag{2.49}$$

Equations (2.48) and (2.49) are shown in Fig. 2.9.

We illustrate these results with the following example.

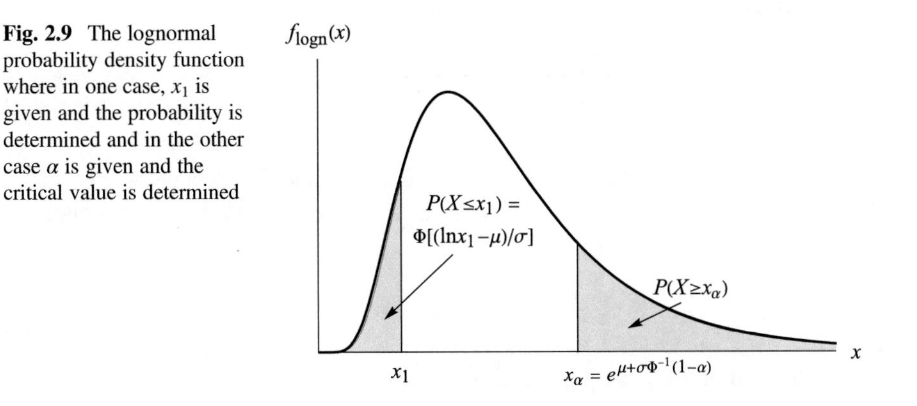

Fig. 2.9 The lognormal probability density function where in one case, x_1 is given and the probability is determined and in the other case α is given and the critical value is determined

Example 2.4
A product's life, in weeks, has a lognormal distribution with $\mu = 5$ and $\sigma = 0.7$. The mean life and variance are obtained from Table 2.1 as

$$E(x) = e^{\mu+\sigma^2/2} = e^{5+0.49/2} = 189.6$$
$$\text{Var}(x) = \left(e^{\sigma^2} - 1\right)e^{2\mu+\sigma^2} = \left(e^{0.49} - 1\right)e^{10+0.49} = 22,734.4$$

and the standard deviation is 150.8.

The probability that the product's life will exceed 250 weeks is, from Eq. (2.48),

$$P(X \geq 250) = 1 - \Phi\left(\frac{\ln(250) - 5}{0.7}\right) = 1 - \Phi(0.6866) = 1 - 0.7538 = 0.2462$$

where we have used interactive graphic IG2–2 to obtain Φ. Thus, about 25% of the products will still be functioning after 250 weeks.

The number of weeks that 95% of the products would have failed is determined from Eq. (2.49) as follows. Since $\alpha = 0.05$,

$$x_{0.05} = e^{5+0.7\Phi^{-1}(0.95)} = e^{5+0.7\times1.645} = 469.4$$

where we have used interactive graphic IG2–2 to obtain Φ^{-1}. Therefore, only 5% of the products will be functioning after 469 weeks.

2.4 Chi Square Distribution

When X_i $i = 1, 2, \ldots, \nu$, are independent standard normal random variables; that is, $X_i \sim N(0,1)$, then, as shown in Sect. A.2 of Appendix A, a chi square random variable X can be generated as

$$X = \sum_{i=1}^{\nu} X_i^2 \sim \chi_\nu^2 \tag{2.50}$$

The quantity ν is called the degrees of freedom and χ_ν^2 denotes a chi square distribution with ν degrees of freedom. It is shown in Sect. A.2 of Appendix A that if

$$Y_1 \sim \chi_{\nu_1}^2, \quad Y_2 \sim \chi_{\nu_2}^2, \quad \ldots \quad , \quad Y_m \sim \chi_{\nu_m}^2$$

then the sum of these random variables

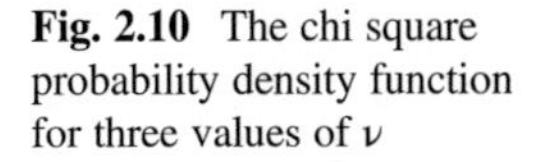

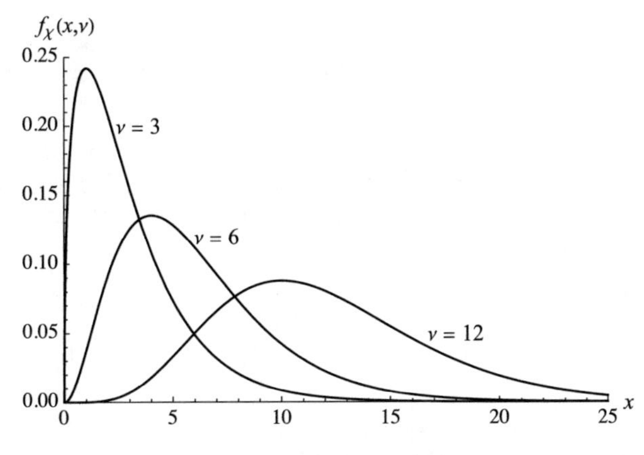

Fig. 2.10 The chi square probability density function for three values of ν

$$Y = Y_1 + Y_2 + \ldots + Y_m \sim \chi^2_{\nu_1+\nu_2+\ldots+\nu_m} \tag{2.51}$$

is also a chi square variable whose degrees of freedom is equal to the sum of the degrees of freedom of the quantities being summed.

From Table 2.1, the probability density function for the chi square distribution is

$$\begin{aligned} f_\chi(x,\nu) &= \frac{2^{-\nu/2}}{\Gamma(\nu/2)} x^{\nu/2-1} e^{-x/2} \quad x > 0 \\ &= 0 \quad x \le 0 \end{aligned} \tag{2.52}$$

which is shown in Fig. 2.10.

Following the notation used in Eq. (2.13), we introduce the following probability statement

$$\Psi_\chi(x,\nu) = P(X \ge x) = \int_x^\infty f_\chi(u,\nu)du \tag{2.53}$$

Consider the case where $\Psi_\chi(x,\nu) = \alpha$. There are two interpretations to this equation. In the first instance, one selects x and the evaluation of the integral results in α. This interpretation will be used in Sect. 2.9.2 to determine a probability called the p-value. In the second instance, one specifies α and then determines that value of $x = \chi_{\alpha,\nu}$ such that the integral evaluates to α. In this case, we are solving the inverse problem

$$\Psi_\chi^{-1}(\alpha) = \chi^2_{\alpha,\nu} \tag{2.54}$$

where the superscript '–1' indicates that the inverse operation has been performed. For this case, $\chi^2_{\alpha,\nu}$ is called the *critical value*. Then, we can write Eq. (2.53) as

$$\Psi_\chi(\chi^2_{\alpha,\nu},\nu) = \alpha = P(X \ge \chi^2_{\alpha,\nu}) = \int_{\chi^2_{\alpha,\nu}}^{\infty} f_\chi(u,\nu)du \tag{2.55}$$

The evaluation of $\Psi_\chi(\chi^2_{\alpha,\nu},\nu) = \alpha$ and $\Psi_\chi^{-1}(\alpha) = \chi^2_{\alpha,\nu}$, which gives the critical value, can be obtained with interactive graphic IG2–2.

We note that

$$\int_{-\infty}^{\infty} f_\chi(u,\nu)du = 1$$

when $\nu > 0$. Therefore, from Eq. (2.55)

$$P(X \le \chi^2_{\alpha,\nu}) = 1 - P(X \ge \chi^2_{\alpha,\nu}) = 1 - \alpha \tag{2.56}$$

and $P(X \le \chi^2_{\alpha,\nu})$ is the cumulative distribution function for chi square. If $\chi^2_{\gamma,\nu} < \chi^2_{\alpha,\nu}$, then

$$P\left(\chi^2_{\gamma,\nu} \le X \le \chi^2_{\alpha,\nu}\right) = P\left(X \ge \chi^2_{\gamma,\nu}\right) - P(X \ge \chi^2_{\alpha,\nu}) = P\left(X \ge \chi^2_{\gamma,\nu}\right) - \alpha$$

Let the probability $P\left(X \le \chi^2_{\gamma,\nu}\right) = \beta$; that is, $\gamma = 1 - \beta$. Then, the above expression can be written as

$$P\left(\chi^2_{1-\beta,\nu} \le X \le \chi^2_{\alpha,\nu}\right) = 1 - \alpha - \beta$$

These results are shown in Fig. 2.11. It is noticed in Fig. 2.11 that as α decreases, $\chi^2_{\alpha,\nu}$ increases and as β decreases $\chi^2_{1-\beta}$ decreases. For the case when $\alpha = \alpha/2$ and $\beta = \alpha/2$, we have that

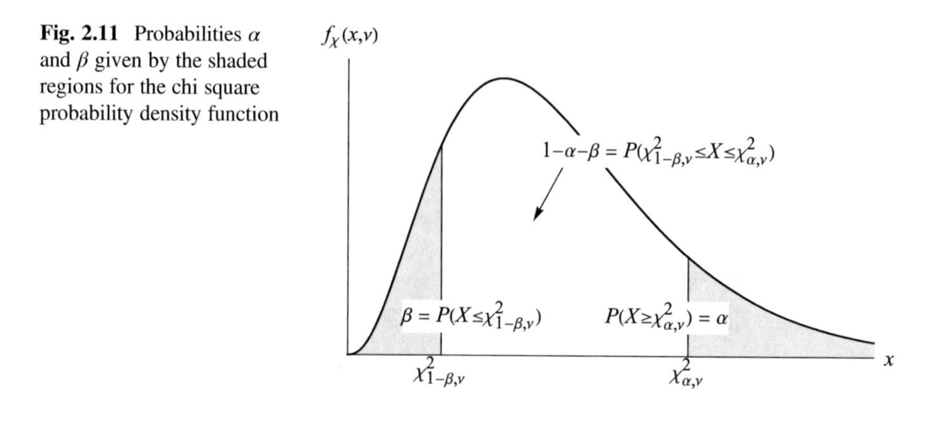

Fig. 2.11 Probabilities α and β given by the shaded regions for the chi square probability density function

$$P\left(\chi^2_{1-\alpha/2,\nu} \le X \le \chi^2_{\alpha/2,\nu}\right) = 1 - \alpha \tag{2.57}$$

We shall now examine the quantity $(n - 1)s^2/\sigma^2$ and show that it is a chi square variable with $n - 1$ degrees of freedom; that is, $(n - 1)s^2/\sigma^2 \sim \chi^2_{n-1}$. In this case, s^2 is the sample variance and σ^2 is the unknown population variance. We start with Eq. (1.38) and divide it by σ^2 to obtain

$$\frac{1}{\sigma^2}\sum_{i=1}^{n}(x_i - \mu)^2 = \frac{1}{\sigma^2}\sum_{i=1}^{n}(x_i - \overline{x})^2 + \frac{n}{\sigma^2}(\overline{x} - \mu)^2 \tag{2.58}$$

Setting

$$Z_i = \frac{x_i - \mu}{\sigma} \quad \text{and} \quad \overline{Z} = \frac{\overline{x} - \mu}{\sigma/\sqrt{n}} \tag{2.59}$$

Eq. (2.58) becomes

$$\sum_{i=1}^{n} Z_i^2 = \frac{(n-1)s^2}{\sigma^2} + \overline{Z}^2 \tag{2.60}$$

Since $Z_i \sim N(0,1)$ and $\overline{Z} \sim N(0,1)$, from Sect. A.2 of Appendix A we see that

$$\begin{aligned} &\sum_{i=1}^{n} Z_i^2 \sim \chi^2_n \\ &\overline{Z}^2 \sim \chi^2_1 \end{aligned} \tag{2.61}$$

are chi square random variables, where the first expression in Eq. (2.61) has a chi square distribution with n degrees of freedom and the second expression is a chi square distribution with one degree of freedom. Then, from Eq. (A.11) of Appendix A,

$$\frac{(n-1)s^2}{\sigma^2} = \sum_{i=1}^{n} Z_i^2 - \overline{Z}^2 \sim \chi^2_{n-1} \tag{2.62}$$

Thus, the quantity $(n - 1)s^2/\sigma^2$ is a chi square random variable with $n - 1$ degrees of freedom. Consequently, in Eq. (2.57),

$$X = \frac{(n-1)s^2}{\sigma^2} \tag{2.63}$$

It is mentioned that when we discuss hypothesis testing in Sect. 2.9, we shall be calling X given by Eq. (2.63) a *test statistic*.

Using Eq. (2.63) in Eq. (2.57), we obtain

$$P\left(\chi^2_{1-\alpha/2,n-1} \le \frac{(n-1)s^2}{\sigma^2} \le \chi^2_{\alpha/2,n-1}\right) = 1 - \alpha$$

which can be written as

$$P\left(\frac{1}{\chi^2_{\alpha/2,n-1}} \le \frac{\sigma^2}{(n-1)s^2} \le \frac{1}{\chi^2_{1-\alpha/2,n-1}}\right) = 1 - \alpha$$

Therefore,

$$P\left(\frac{(n-1)s^2}{\chi^2_{\alpha/2,n-1}} \le \sigma^2 \le \frac{(n-1)s^2}{\chi^2_{1-\alpha/2,n-1}}\right) = 1 - \alpha \tag{2.64}$$

Then, the unknown population variance σ^2 has a $(1 - \alpha)$ probability of being within the limits shown. Thus, we define the $100(1 - \alpha)\%$ *confidence interval* on the population variance σ^2 as

$$\frac{(n-1)s^2}{\chi^2_{\alpha/2,n-1}} \le \sigma^2 \le \frac{(n-1)s^2}{\chi^2_{1-\alpha/2,n-1}} \tag{2.65}$$

This result appears as Case 6 in Table 2.2.

It is noted that the end points of the interval in Eqs. (2.64) and (2.65) are random variables since s is a random variable. Therefore, the interpretation of these expressions is that there is a probability $\alpha/2$ that when a sample is taken from the population $\sigma^2 < (n-1)s^2/\chi^2_{\alpha/2,n-1}$ or a probability $\alpha/2$ that $\sigma^2 > (n-1)s^2/\chi^2_{1-\alpha/2,n-1}$. In other words, if we were to repeat a test with n samples a sufficient number of times, $100(1 - \alpha)\%$ of the time the interval will contain the σ^2.

From Eq. (2.65), we determine that the $100(1 - \alpha)\%$ confidence interval on the upper confidence limit is

$$\sigma^2 \le \frac{(n-1)s^2}{\chi^2_{1-\alpha,n-1}} \tag{2.66}$$

and that on the lower confidence limit is

$$\frac{(n-1)s^2}{\chi^2_{\alpha,n-1}} \le \sigma^2 \tag{2.67}$$

Since the upper and lower bounds are one-sided measures, we have replaced $\chi^2_{1-\alpha/2,n-1}$ with $\chi^2_{1-\alpha,n-1}$ and $\chi^2_{\alpha/2,n-1}$ with $\chi^2_{\alpha,n-1}$.

For specific numerical values for s and n, Eqs. (2.65) to (2.67) can be determined using interactive graphic IG2–3.

If the confidence limits on the standard deviation are of interest, then one takes the positive square root of each term in the above inequalities.

As mentioned previously, as α decreases, the critical value $\chi^2_{\alpha/2,n-1}$ increases and the critical value $\chi^2_{1-\alpha/2,n-1}$ decreases. Therefore, from Eq. (2.65), it is seen that for a given n the effect of decreasing α, which increases the confidence level, is to increase the confidence interval.

The preceding results are illustrated with the following example, which can be verified with interactive graphic IG2–3.

Example 2.5
The measurement of the length of 20 parts that are normally distributed yielded a variance $s^2 = 2.25 \times 10^{-4}$. We shall determine the confidence interval at the 90% confidence level of the population standard deviation σ. Then, from Eq. (2.65),

$$\begin{gathered}\frac{19 \times 2.25 \times 10^{-4}}{\chi^2_{0.05,19}} \le \sigma^2 \le \frac{19 \times 2.25 \times 10^{-4}}{\chi^2_{0.95,19}} \\ \frac{42.75 \times 10^{-4}}{30.14} \le \sigma^2 \le \frac{42.75 \times 10^{-4}}{10.12} \\ 1.418 \times 10^{-4} \le \sigma^2 \le 4.226 \times 10^{-4} \\ 0.0119 \le \sigma \le 0.0206\end{gathered}$$

where the critical values of the chi square distribution were obtained from interactive graphic IG2–2.

2.5 Student t Distribution

The student t distribution (often called simply the t distribution) is obtained when a random variable $X_i \sim N(0,1)$ is divided by an independent random variable that is the square root of a chi square random variable with ν degrees of freedom. The resulting t distribution will have ν degrees of freedom. We know from Eq. (2.62) that

$$\frac{(n-1)s^2}{\sigma^2} \sim \chi^2_{n-1} \tag{2.68}$$

and from Sect. A.3 of Appendix A that for the normal distribution $\bar{x}$, the sample mean, and s^2, the sample variance, are independent. Since $\bar{x} \sim N(\mu, \sigma^2/n)$ then, equivalently, $(\bar{x} - \mu)/\sqrt{\sigma^2/n} \sim N(0, 1)$. Therefore, the statistic

$$T=\frac{\bar{x}-\mu}{\sqrt{s^2/n}}=\frac{(\bar{x}-\mu)/\sqrt{\sigma^2/n}}{\sqrt{((n-1)s^2/\sigma^2)/(n-1)}}\sim\frac{N(0,1)}{\sqrt{\chi^2_\nu/\nu}} \tag{2.69}$$

is a t distribution. In Eq. (2.69), $\nu = n - 1$.

From the central limit theorem, when n is large we expect that $T \sim N(0,1)$. It turns out that there is little difference between the critical values of $t_{\alpha,\nu}$ and z_α for large sample size n. Therefore, one can think of Eq. (2.69) as being used for small sample sizes and using either s or an estimated value for σ when the sample size is large.

When we discuss hypothesis testing in Sect. 2.9, we shall be calling T given by Eq. (2.69) a *test statistic*.

When X_i $i = 1, 2, \ldots, n$ are independent standard normal random variables; that is, $X_i \sim N(0,1)$, we find from Table 2.2 that the probability density function for the t distribution for ν degrees of freedom is

$$f_t(x,\nu)=\frac{1}{\sqrt{\nu}B(\nu/2,1/2)}\left(\frac{\nu}{\nu+x^2}\right)^{(\nu+1)/2} \qquad -\infty\le x\le\infty \tag{2.70}$$

where

$$B(a,b)=\int_0^1 u^{a-1}(1-u)^{b-1}du \tag{2.71}$$

is the beta function and

$$B(\nu/2,1/2)=\frac{\sqrt{\pi}\Gamma(\nu/2)}{\Gamma((\nu+1)/2))}$$

The quantity $\Gamma(x)$ is the gamma function. This distribution is symmetric about $x = 0$ and, therefore, the mean equals the median. The t distribution for two values of ν is shown in Fig. 2.12 along with the standard normal distribution $N(0,1)$.

Following the notation used in Eq. (2.13), we introduce the following probability statement

$$\Psi_t(x,\nu)=P(T\ge x)=\int_x^\infty f_t(u,\nu)du \tag{2.72}$$

Consider the case where $\Psi_t(x,\nu) = \alpha$. There are two interpretations to this equation. In the first instance, one selects x and the evaluation of the integral results in α. This interpretation will be used in Sect. 2.9.2 to determine a probability called the p-value. In the second instance, one specifies α and then determines that value of $x = t_{\alpha,\nu}$ such that the integral evaluates to α. In this case, we are solving the inverse problem

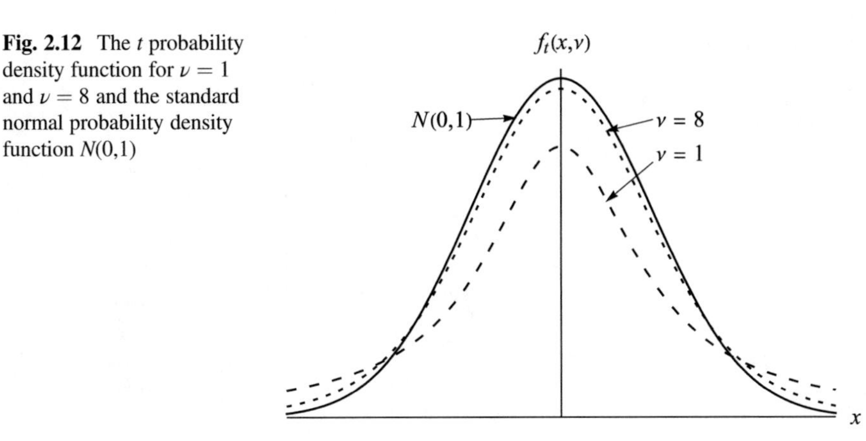

Fig. 2.12 The t probability density function for $\nu = 1$ and $\nu = 8$ and the standard normal probability density function $N(0,1)$

$$\Psi_t^{-1}(\alpha) = t_{\alpha,\nu} \tag{2.73}$$

where the superscript '–1' indicates that the inverse operation has been performed. For this case, $t_{\alpha,\nu}$ is called the *critical value*. Then, we can write Eq. (2.72) as

$$\Psi_t(t_{\alpha,\nu}, \nu) = \alpha = P(T \geq t_{\alpha,\nu}) = \int_{t_{\alpha,\nu}}^{\infty} f_t(u, \nu) du \tag{2.74}$$

The evaluation of $\Psi_t(t_{\alpha,\nu},\nu) = \alpha$ and $\Psi_t^{-1}(\alpha) = t_{\alpha,\nu}$, which gives the critical value, can be obtained with interactive graphic IG2–2.

Since

$$\int_{-\infty}^{\infty} f_t(u, \nu) du = 1$$

we have that

$$P(T \leq t_{\alpha,\nu}) = 1 - P(T \geq t_{\alpha,\nu}) = 1 - \alpha \tag{2.75}$$

and $P(T \leq T_{\alpha,\nu})$ is the cumulative distribution function for the t distribution.

The t distribution is a symmetric distribution about 0; therefore, we can use it in the same manner that was done for the standard normal distribution. Thus, we can use Eq. (2.26) and replace z_α with $t_{\alpha,\nu}$ and z_β with $t_{\beta,\nu}$ to obtain

$$P\left(t_{\beta,\nu} \leq T \leq t_{\alpha,\nu}\right) = 1 - \alpha - \beta \tag{2.76}$$

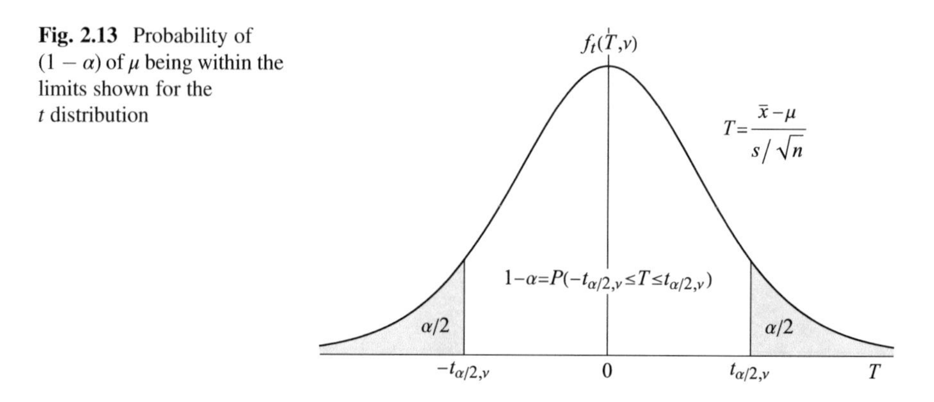

Fig. 2.13 Probability of $(1-\alpha)$ of μ being within the limits shown for the t distribution

where $t_{\beta,\nu} < t_{\alpha,\nu}$ and $\nu = n - 1$. If we set $\beta = \alpha \to \alpha/2$ and $t_{\beta,\nu} = -t_{\alpha/2,\nu}$, then Eq. (2.76) becomes

$$P\left(-t_{\alpha/2,\nu} \le \frac{\bar{x}-\mu}{s/\sqrt{n}} \le t_{\alpha/2,\nu}\right) = 1-\alpha$$

which is shown in Fig. 2.13. This expression can be written as

$$P\left(\bar{x} - t_{\alpha/2,n-1}s/\sqrt{n} \le \mu \le \bar{x} + t_{\alpha/2,n-1}s/\sqrt{n}\right) = 1-\alpha \tag{2.77}$$

which states that the population mean μ has a $(1-\alpha)$ probability of being within the limits shown. Thus, we define the $100(1-\alpha)\%$ *confidence interval* of the population mean μ as

$$\bar{x} - t_{\alpha/2,n-1}s/\sqrt{n} \le \mu \le \bar{x} + t_{\alpha/2,n-1}s/\sqrt{n} \tag{2.78}$$

It is noted that the end points of the interval in Eqs. (2.77) and (2.78) are random variables since s and $\bar{x}$ are random variables. Therefore, the interpretation of these expressions is that there is a probability $\alpha/2$ that when an average is computed from n samples taken from the population $\mu < \bar{x} - t_{\alpha/2,n-1}s/\sqrt{n}$ or a probability $\alpha/2$ that $\mu > \bar{x} + t_{\alpha/2,n-1}s/\sqrt{n}$. In other words, if we were to repeat a test with n samples a sufficient number of times, $100(1-\alpha)\%$ of the time the interval will contain the μ.

From Eq. (2.78), we determine that the $100(1-\alpha)\%$ confidence interval on the upper confidence limit is

$$\mu \le \bar{x} + t_{\alpha,n-1}s/\sqrt{n} \tag{2.79}$$

and that on the lower confidence limit is

$$\overline{x} - t_{\alpha,n-1}s/\sqrt{n} \leq \mu \tag{2.80}$$

Since the upper and lower bounds are one-sided measures, we have replaced $t_{\alpha/2,n\text{-}1}$ with $t_{\alpha,n\text{-}1}$.

The results given by Eqs. (2.78) to (2.80) appear as Case 2 in Table 2.2 and for specific numerical values for s and n they can be determined using interactive graphic IG2–3.

As mentioned previously, $t_{\alpha,\nu}$ behaves in a manner similar to z_α; therefore, it is seen that for a given ν as α decreases $|t_{\alpha,\nu}|$ increases. Therefore, from Eq. (2.78), it is seen that for a given n the effect of decreasing α, which increases the confidence level, is to increase the confidence interval.

When n becomes large (say, $n > 30$), the test statistic given by Eq. (2.69) tends towards a normal distribution as dictated by the central limit theorom. In this case, one may be able to replace $t_{\alpha/2}$ with $z_{\alpha/2}$.

We see from an examination of Eqs. (2.78) to (2.80) that there are two ways to decrease the confidence intervals: increase α or increase n. Since we often want a confidence level that is as high as practical, increasing α is not a good alternative. Therefore, the only option is to increase the number of samples n. For the two-sided confidence interval, the quantity that governs the magnitude of the confidence interval is

$$\Delta_0 = t_{\alpha/2,n-1}\frac{s}{\sqrt{n}} \tag{2.81}$$

If Δ_0 is given a value, then for a given α we choose

$$n = \left\lceil \left(t_{\alpha/2,n-1}\frac{s}{\Delta_0} \right)^2 \right\rceil \tag{2.82}$$

where $\lceil x \rceil$ is the ceiling function indicating that x will become the smallest integer greater than x. Unfortunately, both sides of Eq. (2.82) are a function of n so that a numerical procedure is often required to get a solution. In addition, s is not known before a test is run; therefore, at best one only can get an estimate of n based on a previous run. However, if s varies a 'small' amount from test to test, this value may provide a stable estimate of n.

The preceding results are illustrated with the following example.

Example 2.6

For a set of 23 samples from an assumed normal population, it is found that $\overline{x} = 500$ and $s = 50$. At the 95% confidence level, we shall determine the confidence interval for μ. Thus, from Eq. (2.78) with $\alpha = 0.05$, we find that

$$500 - t_{0.025,22}\frac{50}{\sqrt{23}} \le \mu \le 500 + t_{0.025,22}\frac{50}{\sqrt{23}}$$
$$500 - 2.074\frac{50}{\sqrt{23}} \le \mu \le 500 + 2.074\frac{50}{\sqrt{23}}$$
$$478.4 \le \mu \le 521.6$$

where the critical value for the t distribution was obtained from interactive graphic IG2–2.

This result can be verified with interactive graphic IG2–3.

2.6 Differences in the Means

We extend the results of the previous sections to the case where we can estimate the variability of the difference between the means of two normally distributed independent populations when their variances are either known or unknown and not necessarily equal. We are interested in the quantity $\overline{X}_1 - \overline{X}_2 \sim N(\mu, \sigma^2)$. Then, from Sect. 1.3.6, we find that

$$E(\overline{X}_1 - \overline{X}_2) = E(\overline{X}_1) - E(\overline{X}_2) = \mu_1 - \mu_2 \tag{2.83}$$

which is normally distributed. Since X_1 and X_2 are independent random variables the covariance is zero and, therefore,

$$\mathrm{Var}(\overline{X}_1 - \overline{X}_2) = \mathrm{Var}(\overline{X}_1) + \mathrm{Var}(\overline{X}_2) = \frac{\sigma_1^2}{n_1} + \frac{\sigma_2^2}{n_2} \tag{2.84}$$

Thus,

$$\overline{X}_1 - \overline{X}_2 \sim N(\mu_1 - \mu_2, \sigma_1^2/n_1 + \sigma_2^2/n_2) \tag{2.85}$$

and the form of the standard variable becomes

$$Z = \frac{\overline{X}_1 - \overline{X}_2 - (\mu_1 - \mu_2)}{\sqrt{\sigma_1^2/n_1 + \sigma_2^2/n_2}} \tag{2.86}$$

When the variance σ_1^2 and σ_2^2 are known, the $100(1 - \alpha)\%$ *confidence interval* of the difference of two independent population means $\mu_1 - \mu_2$ is obtained from Eq. (2.36) as

$$\begin{aligned}\overline{X}_1 - \overline{X}_2 - z_{\alpha/2}\sqrt{\sigma_1^2/n_1 + \sigma_2^2/n_2} \le \mu_1 - \mu_2 \le \\ \overline{X}_1 - \overline{X}_2 + z_{\alpha/2}\sqrt{\sigma_1^2/n_1 + \sigma_2^2/n_2}\end{aligned} \tag{2.87}$$

which appears as Case 3 in Table 2.2. When the upper and lower limits are both positive, then one can conclude that $\mu_1 > \mu_2$ and when they are both negative that $\mu_1 < \mu_2$. When the upper and lower limits have different signs, these conclusions cannot be made; that is, μ_1 can be greater than or less than μ_2 with the same probability.

Equation (2.87) can be evaluated numerically with interactive graphic IG2–3.

When n_1 and n_2 are each 'large', we use the central limit theorem and replace σ_1 with s_1 and σ_2 with s_2 in Eq. (2.87). Then, Eq. (2.87) becomes

$$\begin{aligned}\overline{X}_1 - \overline{X}_2 - z_{\alpha/2}\sqrt{s_1^2/n_1 + s_2^2/n_2} \le \mu_1 - \mu_2 \le \\ \overline{X}_1 - \overline{X}_2 + z_{\alpha/2}\sqrt{s_1^2/n_1 + s_2^2/n_2}\end{aligned} \tag{2.88}$$

and the standard variable given by Eq. (2.86) becomes

$$Z = \frac{\overline{X}_1 - \overline{X}_2 - (\mu_1 - \mu_2)}{\sqrt{s_1^2/n_1 + s_2^2/n_2}} \tag{2.89}$$

We now determine the confidence interval when the variances are unknown. This determination depends on whether the variances are unknown and equal or unknown and unequal.

Variances unknown and equal: $\sigma_1^2 = \sigma_2^2 = \sigma^2$.

We assume that one population has a mean μ_1 and variance σ^2 and the other population a mean μ_2 and variance σ^2. From the first population, we use n_1 samples to determine the sample mean $\overline{x}_1$ and sample variance s_1^2 and from the second population we use n_2 samples to determine the sample mean $\overline{x}_2$ and sample variance s_2^2. For this case, we note from Eq. (2.62) that

$$\frac{(n_1 - 1)s_1^2}{\sigma^2} \sim \chi^2_{n_1-1} \quad \text{and} \quad \frac{(n_2 - 1)s_2^2}{\sigma^2} \sim \chi^2_{n_2-1} \tag{2.90}$$

and from Eq. (2.51) that

$$\frac{(n_1 - 1)s_1^2}{\sigma^2} + \frac{(n_2 - 1)s_2^2}{\sigma^2} \sim \chi^2_{n_1+n_2-2} \tag{2.91}$$

is a chi square distribution with $n_1 + n_2 – 2$ degrees of freedom. We set

$$S_p^2 = \frac{(n_1 - 1)s_1^2 + (n_2 - 1)s_2^2}{n_1 + n_2 - 2} \tag{2.92}$$

which is called the *pooled estimator* of σ^2 and when divided by σ^2 has a chi square distribution.

To create the test statistic T, we recall from Eq. (2.69) that the T statistic is the ratio of a statistic with a normal distribution to a statistic that is the square root of a chi square distribution. Therefore, upon using Eq. (2.86) with $\sigma_1 = \sigma_2 = \sigma$ and Eq. (2.92) divided by σ^2, we form the T statistic as follows

$$T = \frac{Z|_{\sigma_1=\sigma_2=\sigma}}{S_p/\sigma} = \frac{\overline{X}_1 - \overline{X}_2 - (\mu_1 - \mu_2)}{S_p\sqrt{1/n_1 + 1/n_2}} \tag{2.93}$$

Then, following the form of Eq. (2.88), the $100(1-\alpha)\%$ *confidence interval* of the difference of two independent population means $\mu_1 - \mu_2$ is

$$\begin{aligned}\overline{X}_1 - \overline{X}_2 - t_{\alpha/2,\nu}S_p\sqrt{1/n_1 + 1/n_2} \le \mu_1 - \mu_2 \le \\ \overline{X}_1 - \overline{X}_2 + t_{\alpha/2,\nu}S_p\sqrt{1/n_1 + 1/n_2}\end{aligned} \tag{2.94}$$

where $\nu = n_1 + n_2 - 2$. This result also appears as Case 4 in Table 2.2. When the upper and lower limits are both positive, then one can conclude that $\mu_1 > \mu_2$ and when they are both negative that $\mu_1 < \mu_2$. When the upper and lower limits have different signs, these conclusions cannot be made; that is, μ_1 can be greater than or less than μ_2 with the same probability.

Equation (2.94) can be evaluated numerically with interactive graphic IG2–3.

It is mentioned that when we discuss hypothesis testing in Sect. 2.9, we shall be calling T given by Eq. (2.93) a *test statistic*.

Variances unknown and unequal: $\sigma_1^2 \neq \sigma_2^2$.

In this case, T can be approximated by

$$T = \frac{\overline{X}_1 - \overline{X}_2 - (\mu_1 - \mu_2)}{\sqrt{s_1^2/n_1 + s_2^2/n_2}} \tag{2.95}$$

with ν degrees of freedom given by the integer value of (Satterthwaite 1946)

$$\nu = \frac{\left(s_1^2/n_1 + s_2^2/n_2\right)^2}{\left(s_1^2/n_1\right)^2/(n_1 - 1) + \left(s_2^2/n_2\right)^2/(n_2 - 1)} \tag{2.96}$$

When we discuss hypothesis testing in Sect. 2.9, we shall be calling T given by Eq. (2.95) a *test statistic*.

Following the form of Eqs. (2.88) and (2.94), the $100(1-\alpha)\%$ *confidence interval* of the difference of two independent population means $\mu_1 - \mu_2$ is

$$\overline{X}_1 - \overline{X}_2 - t_{\alpha/2,\nu}\sqrt{s_1^2/n_1 + s_2^2/n_2} \le \mu_1 - \mu_2 \le \overline{X}_1 - \overline{X}_2 + t_{\alpha/2,\nu}\sqrt{s_1^2/n_1 + s_2^2/n_2} \quad (2.97)$$

where ν is given by Eq. (2.96). This result also appears as Case 5 in Table 2.2. When the upper and lower limits are both positive, one can conclude that $\mu_1 > \mu_2$ and when they are both negative that $\mu_1 < \mu_2$. When the upper and lower limits have different signs, these conclusions cannot be made; that is, μ_1 can be greater than or less than μ_2 with the same probability.

Equation (2.97) can be evaluated numerically with interactive graphic IG2–3.

It is noted that the end points of the interval in Eqs. (2.94) and (2.97) are random variables since s_1, s_2, $\overline{X}_1$, and $\overline{X}_2$ are random variables. The interpretation is the same as that stated for Eq. (2.78).

It is mentioned that Eq. (2.97) produces good results in almost all cases and, since the population variances are rarely known and are rarely equal, this equation is one that is most frequently used.

2.6.1 Paired Samples

There is an important special case of the previous results for the differences in the means, which is called the testing of *paired samples*. In this case, the samples are not assumed to be independent. The paired test assumes that during the test the interfering factor is constant within each pair and that the difference in response will not be affected by the interfering factor. If the interfering factor can interact to complicate the effect on the response, then one should consider a factorial designed experiment, which is discussed in Sect. 4.5. If this assumption is valid, then the paired samples can remove the effect of variabilities in a factor other than the difference between a characteristic being measured in the two populations. Consider the following examples. To determine the ability to detect a certain chemical in a person using a blood sample or a urine sample one would take a blood sample and a urine sample from *one subject* and compare the differences of the results for that subject. This neutralizes the variability among patients. To compare the detection ability between two radar systems, one would use both systems *simultaneously* to detect the same airplane. This neutralizes the effects of atmospheric conditions, airplane type, and airplane flight path. To compare two types of tests to determine the stress in steel beams, one would use both tests on the *same beam specimen*. This would neutralize any differences in material properties and geometry.

If a sample from one population is denoted X_{1i} and that of the second population X_{2i}, then their difference is

$$D_i = X_{1i} - X_{2i}$$

Since the two observations are taken on the same artifact, they are not independent and, from Eq. (1.27) with $\gamma = -1$

$$\mathrm{Var}(D_i) = \mathrm{Var}(X_{1i}) + \mathrm{Var}(X_{2i}) - 2\mathrm{Cov}(X_{1i}, X_{2i})$$

where, because of the homogeneity of the sample artifact, we can expect $\mathrm{Cov}(X_{1i}, X_{2i})$ to be positive and large. Therefore, there will be a reduction in the $\mathrm{Var}(D_i)$ when compared to that where the samples are independent.

If μ_1 is the mean of one population, μ_2 the mean of a second population, and $\mu_D = \mu_1 - \mu_2$, then using Eq. (2.78) the confidence interval is

$$\overline{d} - t_{\alpha/2,n-1}s_D/\sqrt{n} \le \mu_D \le \overline{d} + t_{\alpha/2,n-1}s_D/\sqrt{n} \tag{2.98}$$

where

$$\begin{aligned} \overline{d} &= \frac{1}{n}\sum_{i=1}^{n} D_i \\ s_D^2 &= \frac{1}{n-1}\sum_{i=1}^{n}\left(D_i - \overline{d}\right)^2 \end{aligned} \tag{2.99}$$

Equation (2.98) is another form of Case 2 of Table 2.2.

We illustrate these results with the following examples, all of which can be verified with interactive graphic IG2–3.

Example 2.7

We shall consider data taken from two normally distributed independent populations and determine the confidence interval for the difference in their respective means when their variances are known. For the first population, it is found from 12 randomly selected samples that $\overline{x}_1 = 94.3$ and it is known that $\sigma_1 = 1.51$. From the second population from 10 randomly selected samples it is found that $\overline{x}_1 = 81.5$ and it is known that $\sigma_2 = 2.58$. Then, if we are interested in a confidence level of 95%, we find from Eq. (2.87) that

$$94.3 - 81.5 - 1.96\sqrt{(1.51)^2/12 + (2.58)^2/10} \le \mu_1 - \mu_2 \le \\ 94.3 - 81.5 + 1.96\sqrt{(1.51)^2/12 + (2.58)^2/10}$$

or

$$10.99 \le \mu_1 - \mu_2 \le 14.61$$

where $z_{\alpha/2} = z_{0.025}$ was obtained from Fig. 2.3c. Since the confidence interval of the difference between the means are both positive, we can conclude with a 95% confidence that $\mu_1 > \mu_2$ within the limits shown.

Example 2.8
We shall consider data taken from two normally distributed independent populations. It is found from 10 randomly selected samples from the first population that $\bar{x}_1 = 99.22$ and $s_1 = 6.509$ and from 11 randomly selected samples from the second population that $\bar{x}_2 = 101.3$ and $s_2 = 3.606$. We shall determine the confidence interval for two cases: (a) the variances are unknown and equal; and (b) the variances are unknown and unequal. We are interested in the case for which $\alpha = 0.05$; that is, the 95% confidence level.

For case (a), we find from Eq. (2.92) that

$$S_p = \sqrt{\frac{(10-1)(6.509)^2 + (11-1)(3.606)^2}{10+11-2}} = 5.188$$

and the number of degrees of freedom is $\nu = 10 + 11 - 2 = 19$. Then from Eq. (2.94) and the fact that $t_{0.025,19} = 2.093$ (obtained from interactive graphic IG2–2), the confidence interval is

$$99.22 - 101.3 - 2.093 \times 5.188\sqrt{1/10 + 1/11} \le \mu_1 - \mu_2 \le 99.22 - 101.3 + 2.093 \times 5.188\sqrt{1/10 + 1/11}$$

which reduces to

$$-6.818 \le \mu_1 - \mu_2 \le 2.66$$

Therefore, because the confidence limits have different signs, we conclude that there is no difference between the means.

For case (b), we find from Eq. (2.96) that

$$\nu = \frac{\left((6.509)^2/10 + (3.606)^2/11\right)^2}{\left((6.509)^2/10\right)^2/(10-1) + \left((3.606)^2/11\right)^2/(11-1)} = 13.76$$

which, when rounded to the nearest integer, gives $\nu = 14$. Then, from Eq. (2.97) and the fact that $t_{0.025,14} = 2.145$ (obtained from interactive graphic IG2–2), the confidence interval is

$$99.22 - 101.3 - 2.145\sqrt{(6.509)^2/10 + (3.606)^2/11} \le \mu_1 - \mu_2 \le 99.22 - 101.3 + 2.145\sqrt{(6.509)^2/10 + (3.606)^2/11}$$

which reduces to

$$-7.109 \le \mu_1 - \mu_2 \le 2.949$$

Again, because the confidence limits have different signs, we conclude that there is no difference between the means.

Example 2.9

For 38 paired samples, it was found that $\overline{d} = -2.7$ and $s_D = 3.6$. We shall determine the confidence interval for μ_D at the 99% confidence level. Thus, with $t_{0.005,37} = 2.715$ (obtained from interactive graphic IG2–2), we find from Eq. (2.98) that

$$-2.7 - 2.715 \times 3.6/\sqrt{38} \le \mu_D \le -2.7 + 2.715 \times 3.6/\sqrt{38}$$
$$-4.286 \le \mu_D \le -1.114$$

Since the lower and upper limits of the confidence interval have the same sign, there is a difference in the two experimental conditions that produced the paired samples.

Example 2.10

The annual number of gallons of gasoline consumption of two groups of trucks was as follows. Group 1 had a mean of 2258 with a standard deviation of 1519 while the second group had a mean of 2637 with a standard deviation of 1138. Group 1 had 663 trucks and group 2 had 413 trucks.

We shall determine the confidence interval at the 95% level as follows. Since the sample sizes are every large, we use Eq. (2.88) to obtain

$$2258 - 2637 - 1.96\sqrt{\frac{(1519)^2}{663} + \frac{(1138)^2}{413}} \le \mu_1 - \mu_2 \le$$
$$2258 - 2637 + 1.96\sqrt{\frac{(1519)^2}{663} + \frac{(1138)^2}{413}}$$

or

$$-538.4 \le \mu_1 - \mu_2 \le -219.6$$

and we have used Fig. 2.3c to obtain $z_{0.025} = 1.96$. Since the upper and lower limits are both negative, at the 95% level we conclude that $\mu_2 > \mu_1$ within the limits shown.

2.7 *f* Distribution

If U and V are two independent random variables each having chi square distributions with U having ν_1 degrees of freedom and V having ν_2 degrees of freedom, then the following ratio has an f distribution

$$F \sim \frac{U/\nu_1}{V/\nu_2} \sim \frac{\chi^2_{(1)\nu_1}/\nu_1}{\chi^2_{(2)\nu_2}/\nu_2} \tag{2.100}$$

which, from Table 2.1, has the probability density function

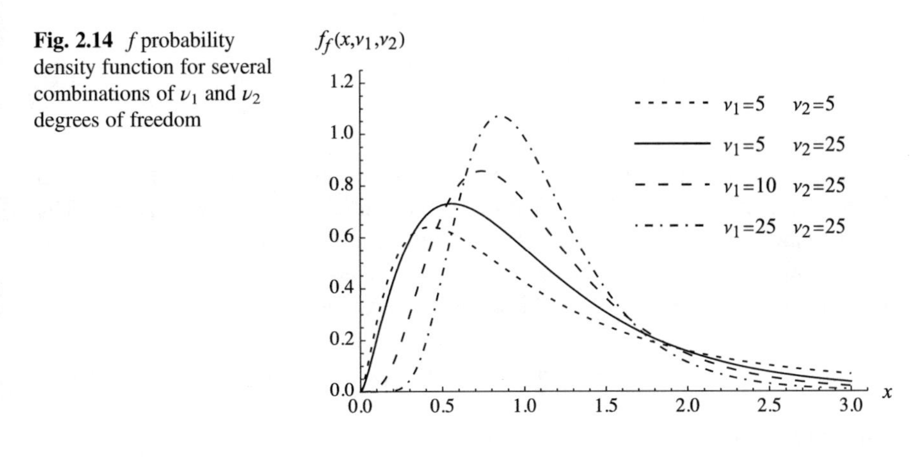

Fig. 2.14 *f* probability density function for several combinations of ν_1 and ν_2 degrees of freedom

$$\begin{aligned} f_f(x,\nu_1,\nu_2) &= \frac{\nu^{\nu_2/2}(\nu_1 x)^{\nu_1/2}(\nu_1 x+\nu_2)^{-(\nu_1+\nu_2)/2}}{xB(\nu_1/2,\nu_2/2)} \quad x>0 \\ &= 0 \quad x \le 0 \end{aligned} \tag{2.101}$$

where $B(a,b)$ is the beta function given by Eq. (2.71). The *f* distribution is shown in Fig. 2.14 for a few combinations of ν_1 and ν_2 degrees of freedom.

Consider two independent normal populations where one population has a variance σ_1^2 and a mean μ_1 and the other has a variance σ_2^2 and a mean μ_2. From the first population, we take n_1 random samples and compute its variance s_1^2 and from the second population we take n_2 random samples and compute its variance s_2^2. Under these assumptions, we know from Eq. (2.62) that the following quantities are chi square random variables

$$\frac{(n_1-1)s_1^2}{\sigma_1^2} \sim \chi^2_{(1)n_1-1} \quad \text{and} \quad \frac{(n_2-1)s_2^2}{\sigma_2^2} \sim \chi^2_{(2)n_2-1} \tag{2.102}$$

Then, substituting Eq. (2.102) into Eq. (2.100), Eq. (2.100) becomes

$$F = \frac{s_1^2/\sigma_1^2}{s_2^2/\sigma_2^2} \tag{2.103}$$

which is an *f* distribution with the numerator having $\nu_1 = n_1–1$ degrees of freedom and the denominator having $\nu_2 = n_2–1$ degrees of freedom. It is mentioned that when we discuss hypothesis testing in Sect. 2.9, we shall be calling F given by Eq. (2.103) a *test statistic*.

Following the notation used in Eq. (2.13), we introduce the following probability statement

$$\Psi_f(x,\nu_1,\nu_2) = P(F \geq x) = \int_x^{\infty} f_f(u,\nu_1,\nu_2)du \tag{2.104}$$

Consider the case where $\Psi_f(x,\nu_1,\nu_2) = \alpha$. There are two interpretations to this equation. In the first instance, one selects x and the evaluation of the integral results in α. This interpretation will be used in Sect. 2.9.2 to determine a probability called the p-value. In the second instance, one specifies α and then determines that value of $x = f_{\alpha,\nu_1,\nu_2}$ such that the integral evaluates to α. In this case, we are solving the inverse problem

$$\Psi_f^{-1}(\alpha) = f_{\alpha,\nu_1,\nu_2} \tag{2.105}$$

where the superscript '–1' indicates that the inverse operation has been performed. For this case, f_{α,ν_1,ν_2} is called the *critical value*. Then, we can write Eq. (2.104) as

$$\Psi_f\left(f_{\alpha,\nu_1,\nu_2},\nu_1,\nu_2\right) = \alpha = P\left(F \geq f_{\alpha,\nu_1,\nu_2}\right) = \int_{f_{\alpha,\nu_1,\nu_2}}^{\infty} f_f(u,\nu_1,\nu_2)du \tag{2.106}$$

The evaluation of $\Psi_f\left(f_{\alpha,\nu_1,\nu_2},\nu_1,\nu_2\right) = \alpha$ and $\Psi_f^{-1}(\alpha) = f_{\alpha,\nu_1,\nu_2}$, which gives the critical value, can be obtained with interactive graphic IG2–2.

Since

$$\int_{-\infty}^{\infty} f_f(u,\nu_1,\nu_2)du = 1$$

we have that

$$P\left(F \leq f_{\alpha,\nu_1,\nu_2}\right) = 1 - P\left(F \geq f_{\alpha,\nu_1,\nu_2}\right) = 1 - \alpha \tag{2.107}$$

and $P\left(F \leq f_{\alpha,\nu_1,\nu_2}\right)$ is the cumulative distribution function for the f distribution. If $f_{\gamma,\nu_1,\nu_2} < f_{\alpha,\nu_1,\nu_2}$, then

$$\begin{aligned} P\left(f_{\gamma,\nu_1,\nu_2} \leq F \leq f_{\alpha,\nu_1,\nu_2}\right) &= P\left(F \geq f_{\gamma,\nu_1,\nu_2}\right) - P\left(F \geq f_{\alpha,\nu_1,\nu_2}\right) \\ &= P\left(F \geq f_{\gamma,\nu_1,\nu_2}\right) - \alpha \end{aligned}$$

Let the probability $P\left(F \leq f_{\gamma,\nu_1,\nu_2}\right) = \beta$; that is, $\gamma = 1 - \beta$. Then the above equation can be written as

$$P\left(f_{1-\beta,\nu_1,\nu_2} \leq F \leq f_{\alpha,\nu_1,\nu_2}\right) = 1 - \alpha - \beta \tag{2.108}$$

These results are shown in Fig. 2.15.

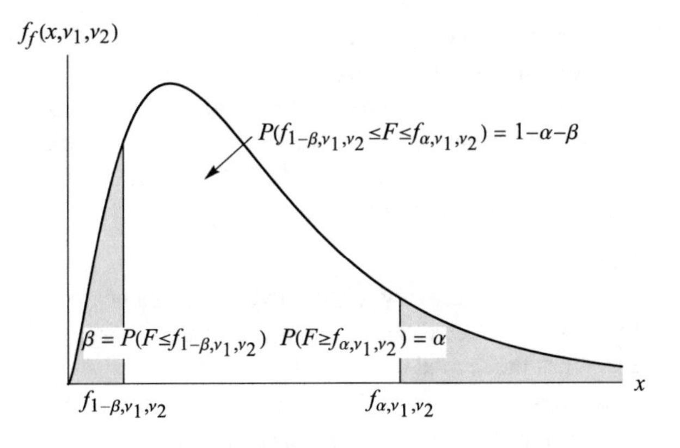

Fig. 2.15 Probabilities α and β given by the shaded regions for the *f* probability density function

Using Eq. (2.103) and setting $\beta = \alpha \rightarrow \alpha/2$, Eq. (2.108) becomes

$$P\left(f_{1-\alpha/2,\nu_1,\nu_2} \frac{s_2^2}{s_1^2} \leq \frac{\sigma_2^2}{\sigma_1^2} \leq f_{\alpha/2,\nu_1,\nu_2} \frac{s_2^2}{s_1^2} \right) = 1 - \alpha \tag{2.109}$$

which states that the ratio of the population variances has a $(1-\alpha)$ probability of being within the limits shown. Thus, we define the $100(1-\alpha)\%$ *confidence interval* on the variance ratio σ_2^2/σ_1^2 of two independent populations as

$$f_{1-\alpha/2,\nu_1,\nu_2} \frac{s_2^2}{s_1^2} \leq \frac{\sigma_2^2}{\sigma_1^2} \leq f_{\alpha/2,\nu_1,\nu_2} \frac{s_2^2}{s_1^2}$$

We now switch to the more conventional notation of placing σ_1 in the numerator, which requires the interchange of the subscripts, and obtain

$$f_{1-\alpha/2,\nu_2,\nu_1} \frac{s_1^2}{s_2^2} \leq \frac{\sigma_1^2}{\sigma_2^2} \leq f_{\alpha/2,\nu_2,\nu_1} \frac{s_1^2}{s_2^2} \tag{2.110}$$

where $\nu_1 = n_1 - 1$ and $\nu_2 = n_2 - 1$. This result appears as Case 7 in Table 2.2. When the upper and lower confidence intervals are less than one, $\sigma_1 < \sigma_2$ and when the upper and lower confidence intervals are greater than one $\sigma_1 > \sigma_2$. When the lower bound is less than one and the upper bound greater than one these conclusions cannot be made; that is, σ_1 can be greater than or less than σ_2 with the same probability.

For specific numerical values for s_1, s_2 ν_1, and ν_2, Eq. (2.110) can be determined using interactive graphic IG2–3.

It is noted that the end points of the interval in Eqs. (2.110) are random variables since s_1 and s_2 are random variables. Therefore, the interpretation of these expressions is that there is a probability $\alpha/2$ that when n_1 and n_2 samples are taken from their respective populations $\sigma_1^2/\sigma_2^2 < f_{1-\alpha/2,\nu_2,\nu_1} s_1^2/s_2^2$ or a probability $\alpha/2$ that $\sigma_1^2/\sigma_2^2 > f_{\alpha/2,\nu_2,\nu_1} s_1^2/s_2^2$. In other words, if we were to repeat the test with the sample number of samples a sufficient number of times, $100(1-\alpha)\%$ of the time the interval will contain the ratio σ_1^2/σ_2^2.

If the confidence limits on the standard deviations are of interest, then one takes the positive square root of each term in the above inequality.

We now illustrate these results with the following example, which can be verified with interactive graphic IG2–3.

Example 2.11
Consider two normally distributed independent processes. From the first process, 16 parts are selected and it is found that the measured characteristic has a variance of 22.09. For the second process, 11 parts are selected and the measured characteristic has a variance of 26.01. We shall determine the confidence interval of σ_1^2/σ_2^2 at the 95% level to determine if there is any difference in the two processes based on their variances. From Eq. (2.110) with $\alpha = 0.05$, $n_1 = 16$, $s_1^2 = 22.09$, $n_2 = 11$, and $s_2^2 = 26.01$, we obtain

$$f_{0.975,10,15}\frac{22.09}{26.01} \le \frac{\sigma_1^2}{\sigma_2^2} \le f_{0.025,10,15}\frac{22.09}{26.01}$$

$$0.284\frac{22.09}{26.01} \le \frac{\sigma_1^2}{\sigma_2^2} \le 3.06\frac{22.09}{26.01}$$

$$0.241 \le \frac{\sigma_1^2}{\sigma_2^2} \le 2.599$$

where the critical values for the f distribution were obtained from interactive graphic IG2–2. Since the lower limit of the ratio of the standard deviations is less than one and the upper limit is greater than one, we cannot state at the 95% confidence level that there is any difference in the processes based on their standard deviations.

2.8 Weibull Distribution

The Weibull distribution is often used to model time-to-failure data, which is used to determine various reliability metrics of components and systems. From Table 2.1, the Weibull probability density function is

$$\begin{aligned} f_W(x) &= \frac{\gamma}{\beta}\left(\frac{x}{\beta}\right)^{\gamma-1} e^{-(x/\beta)^{\gamma}} \quad x > 0 \\ &= 0 \quad x \le 0 \end{aligned} \tag{2.111}$$

where $\gamma > 0$ is a shape parameter and $\beta > 0$ is a scale parameter and, in the reliability community, is called the characteristic life. It is noted that α is a non-dimensional parameter and that β is a dimensional quantity having the units of x. When $\gamma = 1$ and $\beta = 1/\lambda$, Eq. (2.111) reduces to the exponential distribution, which is given in the last row of Table 2.1. When $\gamma = 2$ and $\beta^2 = 2\lambda^2$, Eq. (2.111) reduces to the Rayleigh distribution, which is given in the next to last row of Table 2.1. The Weibull distribution is shown in Fig. 2.16 for a few combinations γ and β. In Fig. 2.16a, the case for $\gamma = 1$ and $\beta = 1$ corresponds to that of an exponential distribution with $\lambda = 1$ and the case for $\gamma = 2$ and $\beta = 1$ corresponds to that of a Rayleigh distribution with $\lambda = 1/\sqrt{2}$.

The cumulative distribution for the Weibull distribution is, from Table 2.1,

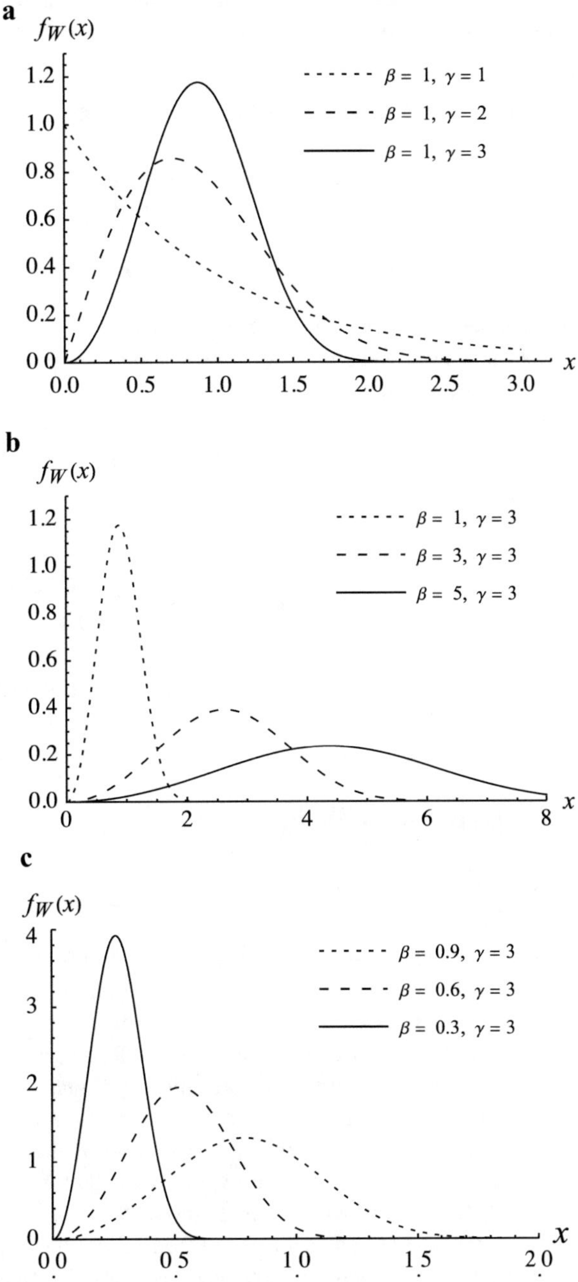

Fig. 2.16 Weibull distribution for the indicated combinations of γ and β (**a**) $\beta = 1$ and $\gamma = 1, 2, 3$ (**b**) $\gamma = 3$ and $\beta = 1, 3, 5$ (**c**) $\gamma = 3$ and $\beta = 0.9, 0.6, 0.3$

$$P(X \leq x) = F_W(x) = 1 - e^{-(x/\beta)^\gamma} \tag{2.112}$$

We shall illustrate these results with the following example.

Example 2.12
In measuring product life, it was found that it had a Weibull distribution with $\gamma = 3$ and $\beta = 8000$ hours. From Table 2.1, the expected value and variance are

$$E(x) = \beta\Gamma\left(1 + \frac{1}{\gamma}\right) = 8000\Gamma(4/3) = 7144$$

$$\text{Var}(x) = \beta^2\left(\Gamma\left(1 + \frac{2}{\gamma}\right) - \Gamma\left(1 + \frac{1}{\gamma}\right)^2\right) = 64 \times 10^6\left(\Gamma(5/3) - [\Gamma(4/3)]^2\right)$$
$$= 6.741 \times 10^6$$

and the standard deviation is 2596.4 hours.

The probability that this product will last more than 11,000 hours is obtained from Eq. (2.106), which can be written as

$$P(X \geq x) = 1 - P(X \leq x) = 1 - F_W(x) = e^{-(x/\beta)^\gamma}$$

Therefore,

$$P(X \geq 11000) = \exp\left(-\left(\frac{11000}{8000}\right)^3\right) = 0.0743$$

Thus, about 7.4% of the products' lives will exceed 11,000 hours.

2.9 Hypothesis Testing

2.9.1 Introduction

A *statistical hypothesis* is a statement about a measurable parameter of one or more populations. *Hypothesis testing* assesses the plausibility of the hypothesis by using randomly selected data from these populations. The correctness of the hypothesis is never known with certainty unless the entire population is examined. Therefore, there is always a possibility of arriving at an incorrect conclusion. Since we use probability distributions to represent populations, a statistical hypothesis implies an assumption about the probability distribution of the sample. The language that is used when one tests a statistical hypothesis is that one either "rejects it" or "fails to reject it". A difficulty with saying that one "accepts" a hypothesis is that this seems to imply that the hypothesis is true when in fact the hypothesis always has a finite probability of being wrong. However, "accepts" is often used.

We now formalize the hypothesis test process by introducing a statistical quantity of interest denoted θ. This statistic can be, for example, the mean of a population, the difference in the means of two populations, or the ratio of the variances of two populations. This statistic may depend on whether the variance of the population is known or unknown as indicated, for example, by Eqs. (2.88) and (2.94). Therefore, we shall set up the general procedure in terms of θ and then illustrate its use for different statistical quantities. In all cases, the hypothesis decision is based on the mean or the variance determined from an appropriate number of samples taken from the population(s).

The procedure involves stating a hypothesis called the null hypothesis, denoted H_0, and an alternative hypothesis, denoted H_A, in the following manner. If the test is to be a *two-sided* test, then the statement reads

$$H_0: \quad \theta = \theta_0 \text{ or } H_A: \quad \theta \neq \theta_0 \tag{2.113}$$

It is considered two-sided because the expression $\theta \neq \theta_0$ implies that θ can be either greater than or less than θ_0. If the test is to be a *one-sided* test, then the statement reads as either

$$H_0: \quad \theta \leq \theta_0 \text{ or } H_A: \quad \theta > \theta_0 \tag{2.114}$$

or

$$H_0: \quad \theta \geq \theta_0 \text{ or } H_A: \quad \theta < \theta_0 \tag{2.115}$$

As will be shown subsequently, implied in these statements is that the decision to reject or to fail to reject the null hypothesis is a decision always made at some level of significance in an analogous manner that was used to determine confidence levels. The significance level, denoted α, is the probability of rejecting the null hypothesis when it is true. If we fail to reject H_0, this means that H_0 is a plausible statement, but it does not mean that H_0 has been proven true. In fact, H_0 cannot be proven to be true; it can only be shown to be implausible.

We see from this discussion that it is possible to reject H_0 when it is true. Such an error is called a type I error. The type I error is also called the α error, the significance level, or the size of the test. Conversely, it is possible to fail to reject H_0 when it is false. This error is called a type II error or the β error. The β error is used to define a quantity called the power, which equals $1 - \beta$ and can be thought of as the probability of correctly rejecting a false H_0. In all cases, one would like to keep α and β small. The definitions of the type I and type II errors are given in Sect. 2.9.4.

Before proceeding, we note from the material in the previous sections that the probability statements are typically of the form

$$P\left(\theta_\beta \leq \Theta_0 \leq \theta_\alpha\right) = 1 - \alpha - \beta \tag{2.116}$$

where Θ_0 is the test statistic given in Table 2.3 and θ_α and θ_β are the corresponding critical values. Equation (2.116) states the probability that Θ_0 lies between these

Table 2.3 Hypothesis test criteria for various statistical quantities,[a] where $\delta_0 = \mu_1 - \mu_2$, $\nu = n - 1$, $\nu_1 = n_1 - 1$, $\nu_2 = n_2 - 1$, and $\nu_4 = n_1 + n_2 - 2$

Case	Statistic	Assumptions	Test statistic (Θ_0)	Null hypothesis	Alternate hypotheses	Rejection criteria[b]	p-value
1	μ	σ^2 (known)	$Z_0 = \dfrac{\overline{x} - \mu_0}{\sigma/\sqrt{n}}$	$H_0 : \mu = \mu_0$	$H_A : \mu \neq \mu_0$ $H_A : \mu > \mu_0$ $H_A : \mu < \mu_0$	$Z_0 < -z_{\alpha/2}$ or $Z_0 > z_{\alpha/2}$ $Z_0 > z_\alpha$ $Z_0 < -z_\alpha$	$2\Psi_N(\lvert Z_0\rvert)$ $\Psi_N(\lvert Z_0\rvert)$ $\Psi_N(\lvert Z_0\rvert)$
2	μ	σ^2 (unknown)	$T_0 = \dfrac{\overline{x} - \mu_0}{s/\sqrt{n}}$	$H_0 : \mu = \mu_0$	$H_A : \mu \neq \mu_0$ $H_A : \mu > \mu_0$ $H_A : \mu < \mu_0$	$T_0 < -t_{\alpha/2,\nu}$ or $T_0 > t_{\alpha/2,\nu}$ $T_0 > t_{\alpha,\nu}$ $T_0 < -t_{\alpha,\nu}$	$2\Psi_t(\lvert T_0\rvert, \nu)$ $\Psi_t(\lvert T_0\rvert, \nu)$ $\Psi_t(\lvert T_0\rvert, \nu)$
3	$\mu_1 - \mu_2$	σ_1^2, σ_2^2 (known)	$Z_0 = \dfrac{\overline{x}_1 - \overline{x}_2 - \delta_0}{\sqrt{\sigma_1^2/n_1 + \sigma_2^2/n_2}}$	$H_0 : \mu_1 - \mu_2 = \delta_0$	$H_A : \mu_1 - \mu_2 \neq \delta_0$ $H_A : \mu_1 - \mu_2 > \delta_0$ $H_A : \mu_1 - \mu_2 < \delta_0$	$Z_0 < -z_{\alpha/2}$ or $Z_0 > z_{\alpha/2}$ $Z_0 > z_\alpha$ $Z_0 < -z_\alpha$	$2\Psi_N(\lvert Z_0\rvert)$ $\Psi_N(\lvert Z_0\rvert)$ $\Psi_N(\lvert Z_0\rvert)$
4	$\mu_1 - \mu_2$	$\sigma_1^2 = \sigma_2^2 = \sigma^2$ (unknown)	$T_0 = \dfrac{\overline{x}_1 - \overline{x}_2 - \delta_0}{S_p\sqrt{1/n_1 + 1/n_2}}$ S_p from Eq.(2.92)	$H_0 : \mu_1 - \mu_2 = \delta_0$	$H_A : \mu_1 - \mu_2 \neq \delta_0$ $H_A : \mu_1 - \mu_2 > \delta_0$ $H_A : \mu_1 - \mu_2 < \delta_0$	$T_0 < -t_{\alpha/2,\nu_4}$ or $T_0 > t_{\alpha/2,\nu_4}$ $T_0 > t_{\alpha,\nu_4}$ $T_0 < -t_{\alpha,\nu_4}$	$2\Psi_t(\lvert T_0\rvert, \nu_4)$ $\Psi_t(\lvert T_0\rvert, \nu_4)$ $\Psi_t(\lvert T_0\rvert, \nu_4)$
5	$\mu_1 - \mu_2$	$\sigma_1^2 \neq \sigma_2^2$ (unknown)	$T_0 = \dfrac{\overline{x}_1 - \overline{x}_2 - \delta_0}{\sqrt{s_1^2/n_1 + s_2^2/n_2}}$ ν_5 from Eq.(2.96)	$H_0 : \mu_1 - \mu_2 = \delta_0$	$H_A : \mu_1 - \mu_2 \neq \delta_0$ $H_A : \mu_1 - \mu_2 > \delta_0$ $H_A : \mu_1 - \mu_2 < \delta_0$	$T_0 < -t_{\alpha/2,\nu_5}$ or $T_0 > t_{\alpha/2,\nu_5}$ $T_0 > t_{\alpha,\nu_5}$ $T_0 < -t_{\alpha,\nu_5}$	$2\Psi_t(\lvert T_0\rvert, \nu_5)$ $\Psi_t(\lvert T_0\rvert, \nu_5)$ $\Psi_t(\lvert T_0\rvert, \nu_5)$

6	σ^2	$X_0 = \dfrac{(n-1)s^2}{\sigma_0^2}$	$H_0: \sigma^2 = \sigma_0^2$	$H_A: \sigma^2 \neq \sigma_0^2$	$X_0 < \chi^2_{1-\alpha/2,\nu}$ or $X_0 > \chi^2_{\alpha/2,\nu}$	$2\Psi_\chi(X_0, \nu)$
				$H_A: \sigma^2 > \sigma_0^2$	$X_0 > \chi^2_{\alpha,\nu}$	$\Psi_\chi(X_0, \nu)$
				$H_A: \sigma^2 < \sigma_0^2$	$X_0 < \chi^2_{1-\alpha,\nu}$	$\Psi_\chi(X_0, \nu)$
7	$\dfrac{\sigma_1^2}{\sigma_2^2}$	$F_0 = \dfrac{s_1^2}{s_2^2}$	$H_0: \sigma_1^2 = \sigma_2^2$	$H_A: \sigma_1^2 \neq \sigma_2^2$	$F_0 < f_{1-\alpha/2,\nu_1,\nu_2}$ or $F_0 > f_{\alpha/2,\nu_1,\nu_2}$	$2\Psi_f(F_0, \nu_1, \nu_2)$
				$H_A: \sigma_1^2 > \sigma_2^2$	$F_0 > f_{\alpha,\nu_1,\nu_2}$	$\Psi_f(F_0, \nu_1, \nu_2)$
				$H_A: \sigma_1^2 < \sigma_2^2$	$F_0 < f_{1-\alpha,\nu_1,\nu_2}$	$\Psi_f(F_0, \nu_1, \nu_2)$

[a]The expressions in this table can be numerically evaluated with interactive graphic IG2–4
[b]Numerical values for the critical values and the p-values can be determined from interactive graphic IG2-2

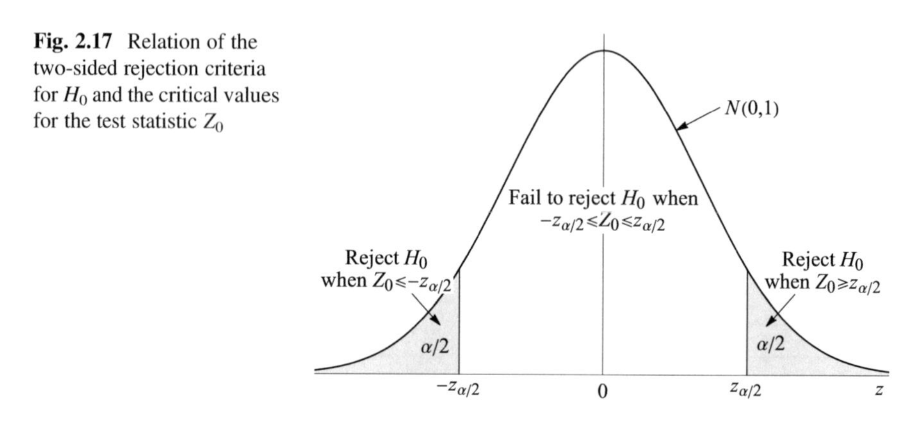

Fig. 2.17 Relation of the two-sided rejection criteria for H_0 and the critical values for the test statistic Z_0

critical values. Therefore, when Θ_0 lies outside these values, we can use this fact to conduct a hypothesis test as discussed next.

There are two ways in which to decide if H_0 is plausible. Both ways are summarized in Table 2.3, which has its genesis in Table 2.2 and the confidence intervals determined in Sect. 2.3 to 2.6. The first way is to state a significance level α and proceed as follows:

1. State H_0 and H_A.
2. For the given data, compute the appropriate test statistic as listed in the 'Test statistic' column of Table 2.3.
3. For this test statistic, determine the appropriate critical value(s) for α as indicated in the footnote of Table 2.3.
4. Compare the test statistic to the critical value and based on their relative magnitudes as indicated in the column labeled 'Rejection criteria' in Table 2.3 determine the plausibility of H_0.

This procedure is illustrated in Fig. 2.17 for the test statistic Z_0.

For a given α, the critical values in step 4 are determined from the following equations: (i) normal distribution, Eq. (2.15) which yields z_α; (ii) chi square distribution, Eq. (2.54) which yields $\chi^2_{\alpha,\nu}$; (iii) t distribution, Eq. (2.73) which yields $t_{\alpha,\nu}$; and iv) f ratio distribution, Eq. (2.105) which yields f_{α,ν_1,ν_2}. Numerical values for all of these quantities can be obtained from interactive graphic IG2–2.

In Sect. 2.9.3, this method is illustrated for the cases presented in Table 2.3.

The second way involves determining a quantity called the p-value, which is discussed next.

2.9.2 p-*Value*

The determination of a quantity called the p-value provides another value by which a decision to reject H_0 can be made. Instead of assuming a value for the significance level α and determining the critical value, one computes the test statistic Θ_0 and based on its value determines the corresponding value for the probability, which is denoted p. Then, based on the magnitude of p a decision is made on whether to reject H_0.

The p-value is determined by using the appropriate test statistic given in Table 2.3 in the following manner. For the normal distribution, the test statistic is $|Z_0|$ and we use Eq. (2.13) to obtain the p-value from

$$p\text{-value} = \Psi_N(|Z_0|) = \int_{|Z_0|}^{\infty} f_N(u)du \tag{2.117}$$

For the chi square distribution, the test statistic is X_0 and we use Eq. (2.55) to obtain the p-value from

$$p\text{-value} = \Psi_\chi(X_0, \nu) = \int_{X_0}^{\infty} f_\chi(u, \nu)du \tag{2.118}$$

For the t distribution, the test statistic is $|T_0|$ and we use Eq. (2.72) to obtain the p-value from

$$p\text{-value} = \Psi_t(|T_0|, \nu) = \int_{|T_0|}^{\infty} f_t(u, \nu)du \tag{2.119}$$

For the f distribution, the test statistic is F_0 and we use Eq. (2.104) to obtain the p-value from

$$p\text{-value} = \Psi_f(F_0, \nu_1, \nu_2) = \int_{F_0}^{\infty} f_f(u, \nu_1, \nu_2)du \tag{2.120}$$

Numerical values for Eqs. (2.117) to (2.120) can be obtained with interactive graphic IG2–2.

Although there are no hard rules, a p-value < 0.01 is one in which H_0 is not considered a plausible statement; that is, we reject the null hypothesis. On the other hand, a p-value > 0.1 is one in which H_0 is considered a plausible statement and, therefore, the null hypothesis is not rejected. When the p-value falls in the region $0.01 < p\text{-value} < 0.1$ there is some evidence that the null hypothesis is not plausible, but the evidence is not strong, and the results tend to be less conclusive. In any event, irrespective of the p-value criterion, the smaller the p-value becomes the less plausible is the null hypothesis. In essence, we have 'replaced' the pre-selected value for α with p with the distinction that we did not chose p, instead p is determined by the data and the hypothesis. We simply interpret its magnitude. In other words, the p-value *is the observed significance level*. Therefore, the p-value provides a means to decide about H_0 based the magnitude of the observed significance level rather than on a preselected level. When used in conjunction with the fixed value α, the p-value provides the additional information as to the degree to which p is less than α. The smaller it is with respect to α the more confidant one can be in rejecting H_0.

To summarize the p-value method:

1. State H_0 and H_A.
2. For the given data, compute the appropriate test statistic as listed in the 'Test statistic' column of Table 2.3.
3. For this test statistic, determine the p-value as indicated in the last column of Table 2.3.
4. Based on the p-value make a judgement about the plausibility of H_0.

Examples of how the p-value is used are given in the next section.

2.9.3 Examples of Hypothesis Testing

In this section, we illustrate hypothesis testing discussed in Sect. 2.9.1 and 2.9.2 for each case appearing in Table 2.3.

Example 2.13 Case 1 of Table 2.3

The mean of the population should be 50.2. A random selection of 21 samples yields an average $\overline{x} = 49.2$. It is known that the standard deviation of the population is $\sigma = 1.5$. Based on this sample, we would like to determine if the requirement that $\mu = 50.2$ is being met at the 98% confidence level.

The null hypothesis statement is H_0: $\mu = 50.2$ with the alternative hypothesis being H_A: $\mu \neq 50.2$. Thus, $\mu_0 = 50.2$ and the test statistic is

$$Z_0 = \frac{\overline{x} - \mu_0}{\sigma/\sqrt{n}} = \frac{49.2 - 50.2}{1.5/\sqrt{21}} = -3.055$$

For $\alpha = 0.02$, we find from Fig. 2.3b that $-z_{0.01} = z_{0.99} = 2.326$. From the last column in Table 2.3 and Eqs. (2.117) and (2.19), the p-value is

$$p - \text{value} = 2\Psi_N(|Z_0|) = 2(1 - \Phi(|Z_0|)) = 2(1 - 0.99888) = 0.00225$$

where Φ was obtained from interactive graphic IG2–2. Since Z_0 $(= -3.055) < z_{0.01}$ $(= -2.326)$ we satisfy the rejection criteria given in Table 2.3. In addition, the p-value < 0.01, which also indicates that we can reject the null hypothesis and favor the alternative hypothesis.

The verification of these results is obtained with interactive graphic IG2–4, which yields Table 2.4.

Table 2.4 Hypothesis test results using interactive graphic IG2–4 for Example 2.13

$Z_0 = -3.055$ H_0: $\mu = \mu_0 = 50.2$	Rejection criteria		
Confidence level	H_A: $\mu \neq \mu_0$	H_A: $\mu > \mu_0$	H_A: $\mu < \mu_0$
90% ($\alpha = 0.1$)	$Z_0 < -1.645$ or $Z_0 > 1.645$	$Z_0 > 1.282$	$Z_0 < -1.282$
95% ($\alpha = 0.05$)	$Z_0 < -1.96$ or $Z_0 > 1.96$	$Z_0 > 1.645$	$Z_0 < -1.645$
99% ($\alpha = 0.01$)	$Z_0 < -2.576$ or $Z_0 > 2.576$	$Z_0 > 2.326$	$Z_0 < -2.326$
99.9% ($\alpha = 0.001$)	$Z_0 < -3.291$ or $Z_0 > 3.291$	$Z_0 > 3.09$	$Z_0 < -3.09$
p-value	0.00225	0.001125	0.001125

Table 2.5 Hypothesis test results using interactive graphic IG2–4 for Example 2.14

$T_0 = 3.2$ H_0: $\mu = \mu_0 = 1.64$	Rejection criteria		
Confidence level	H_A: $\mu \neq \mu_0$	H_A: $\mu > \mu_0$	H_A: $\mu < \mu_0$
90% ($\alpha = 0.1$)	$T_0 < -1.753$ or $T_0 > 1.753$	$T_0 > 1.341$	$T_0 < -1.341$
95% ($\alpha = 0.05$)	$T_0 < -2.131$ or $T_0 > 2.131$	$T_0 > 1.753$	$T_0 < -1.753$
99% ($\alpha = 0.01$)	$T_0 < -2.947$ or $T_0 > 2.947$	$T_0 > 2.602$	$T_0 < -2.602$
99.9% ($\alpha = 0.001$)	$T_0 < -4.073$ or $T_0 > 4.073$	$T_0 > 3.733$	$T_0 < -3.733$
p-value	0.005964	0.002982	0.002982

Example 2.14 Case 2 of Table 2.3

The mean of a population should exceed 1.64. A random selection of 16 samples yields an average $\overline{x} = 1.68$ and a standard deviation $s = 0.05$. Based on this sample, we would like to determine if the requirement that $\mu > 1.64$ is being met. We shall use the p-value to determine if this is so.

Thus, $\mu_0 = 1.64$ and the null hypothesis statement is H_0: $\mu = 1.64$ with the alternative hypothesis being H_A: $\mu > 1.64$. Since the variance is unknown, we use the test statistic

$$T_0 = \frac{\overline{x} - \mu_0}{s/\sqrt{n}} = \frac{1.68 - 1.64}{0.05/\sqrt{16}} = 3.2$$

The p-value is determined from Eq. (2.119) as

$$p - \text{value} = \Psi_t(3.2, 15) = 0.00298$$

where the probability Ψ_t was obtained from interactive graphic IG2–2. Since the p-value is less than 0.01, we reject the null hypothesis and conclude that the criterion $\mu > 1.64$ is being met.

The verification of these results is obtained with interactive graphic IG2–4, which yields Table 2.5.

Example 2.15 Case 3 of Table 2.3
Two independent processes are to be compared. The standard deviation of each process is known to be 8.3; that is, $\sigma_1 = \sigma_2 = 8.3$, and the process variables are normally distributed. Random samples are taken from each process with each set containing 11 samples: thus, $n_1 = n_2 = 11$. The average of the sample set from the first process is $\overline{x}_1 = 120$ and that of the second process is $\overline{x}_2 = 110$. We would like to determine if there is any difference between the means of the two processes and, in particular, if the second process causes a reduction in the average value compared to the first process.

The null hypothesis statement is H_0: $\mu_1 - \mu_2 = 0$ with the alternative hypothesis being H_A: $\mu_1 > \mu_2$. Thus, $\delta_0 = 0$ and the test statistic is

$$Z_0 = \frac{\overline{x}_1 - \overline{x}_2 - \delta_0}{\sqrt{\sigma_1^2/n_1 + \sigma_2^2/n_2}} = \frac{120 - 110 - 0}{\sqrt{(8.3)^2/11 + (8.3)^2/11}} = 2.826$$

The p-value is obtained from Eqs. (2.117) and (2.19) as

$$p - \text{value} = \Psi_N(2.826) = 1 - \Phi(2.826) = 1 - 0.9976 = 0.00236$$

The value for Φ was obtained from interactive graphic IG2–2.

Since the p-value is less than 0.01, we reject the null hypothesis and conclude that the second process does reduce the average value of this parameter.

The verification of these results is obtained with interactive graphic IG2–4, which yields Table 2.6.

Example 2.16 Case 4 of Table 2.3
Two independent processes are being compared. A set of random samples is taken from each process with each set containing 9 samples: thus, $n_1 = n_2 = 9$. The first process yields $\overline{x}_1 = 93.3$ and $s_1 = 2.42$ and second process yields $\overline{x}_2 = 93.8$ and $s_2 = 3.01$. We would like to determine if there is any difference between the means of the two processes assuming that the samples from each process are normally distributed and that the variances are unknown and equal.

Table 2.6 Hypothesis test results using interactive graphic IG2–4 for Example 2.15

$Z_0 = 2.826$ H_0: $\mu_1 - \mu_2 = \delta_0 = 0.0$	Rejection criteria		
Confidence level	H_A: $\mu_1 - \mu_2 \neq \delta_0$	H_A: $\mu_1 - \mu_2 > \delta_0$	H_A: $\mu_1 - \mu_2 < \delta_0$
90% ($\alpha = 0.1$)	$Z_0 < -1.645$ or $Z_0 > 1.645$	$Z_0 > 1.282$	$Z_0 < -1.282$
95% ($\alpha = 0.05$)	$Z_0 < -1.96$ or $Z_0 > 1.96$	$Z_0 > 1.645$	$Z_0 < -1.645$
99% ($\alpha = 0.01$)	$Z_0 < -2.576$ or $Z_0 > 2.576$	$Z_0 > 2.326$	$Z_0 < -2.326$
99.9% ($\alpha = 0.001$)	$Z_0 < -3.291$ or $Z_0 > 3.291$	$Z_0 > 3.09$	$Z_0 < -3.09$
p-value	0.00472	0.00236	0.00236

The null hypothesis statement is H_0: $\mu_1 - \mu_2 = 0$ with the alternative hypothesis being H_A: $\mu_1 \neq \mu_2$. From Eq. (2.92),

$$S_p = \sqrt{\frac{(n_1-1)s_1^2 + (n_2-1)s_2^2}{n_1+n_2-2}} = \sqrt{\frac{(9-1)(2.42)^2 + (9-1)(3.01)^2}{9+9-2}} = 2.731$$

Then, the test statistic with $\delta_0 = 0$ is

$$T_0 = \frac{\bar{x}_1 - \bar{x}_2 - \delta_0}{S_p\sqrt{1/n_1 + 1/n_2}} = \frac{93.3 - 93.8 - 0}{2.736\sqrt{1/9 + 1/9}} = -0.3876$$

The p-value is determined from Eq. (2.118) as

$$p - \text{value} = 2\Psi_t(0.3876, 16) = 2(0.3516) = 0.7032$$

where the value for the probability Ψ_t was obtained from interactive graphic IG2–2. Since the p-value is greater than 0.1, we cannot reject the null hypothesis and conclude that the second process does not differ from the first process.

The verification of these results is obtained with interactive graphic IG2–4, which yields Table 2.7.

Example 2.17 Case 5 of Table 2.3

Two independent processes are to be compared. A set of random samples is taken from each process, and it is found for the first process using 11 samples that $\bar{x}_1 = 14.1$ and $s_1 = 7.6$ and for the second process using 12 samples that $\bar{x}_2 = 23.2$ and $s_2 = 14.0$. We assume that the data are normally distributed and that the variances are unknown and unequal. We would like to determine if there is any difference between the means of the two processes.

The null hypothesis statement is H_0: $\mu_1 - \mu_2 = 0$ with the alternative hypothesis being H_A: $\mu_1 \neq \mu_2$. The number of degrees of freedom is determined from Eq. (2.92) as

Table 2.7 Hypothesis test results using interactive graphic IG2–4 for Example 2.16

$T_0 = -0.3884$ H_0: $\mu_1 - \mu_2 = \delta_0 = 0.0$	Rejection criteria		
Confidence level	H_A: $\mu_1 - \mu_2 \neq \delta_0$	H_A: $\mu_1 - \mu_2 > \delta_0$	H_A: $\mu_1 - \mu_2 < \delta_0$
90% ($\alpha = 0.1$)	$T_0 < -1.746$ or $T_0 > 1.746$	$T_0 > 1.337$	$T_0 < -1.337$
95% ($\alpha = 0.05$)	$T_0 < -2.12$ or $T_0 > 2.12$	$T_0 > 1.746$	$T_0 < -1.746$
99% ($\alpha = 0.01$)	$T_0 < -2.921$ or $T_0 > 2.921$	$T_0 > 2.583$	$T_0 < -2.583$
99.9% ($\alpha = 0.001$)	$T_0 < -4.015$ or $T_0 > 4.015$	$T_0 > 3.686$	$T_0 < -3.686$
p-value	0.7032	0.3516	0.3516

$$\begin{aligned}\nu &= \frac{\left(s_1^2/n_1 + s_2^2/n_2\right)^2}{\left(s_1^2/n_1\right)^2/(n_1-1) + \left(s_2^2/n_2\right)^2/(n_2-1)} \\ &= \frac{\left((7.6)^2/11 + (14)^2/12\right)^2}{\left((7.6)^2/11\right)^2/(11-1) + \left((14)^2/12\right)^2/(12-1)} \\ &= 17.25\end{aligned}$$

which, when rounded to the nearest integer, gives $\nu = 17$. With $\delta_0 = 0$, the test statistic is

$$T_0 = \frac{\bar{x}_1 - \bar{x}_2 - \delta_0}{\sqrt{s_1^2/n_1 + s_2^2/n_2}} = \frac{14.1 - 23.2 - 0}{\sqrt{(7.6)^2/11 + (14)^2/12}} = -1.959$$

The p-value is determined from Eq. (2.119) as

$$p - \text{value} = 2\Psi_t(1.959, 17) = 2(0.03326) = 0.06676$$

The value for the probability Ψ_t was obtained from interactive graphic IG2–2.

The p-value falls in the grey area of $0.01 < p < 0.1$. If we had selected an $\alpha = 0.05$ (a 95% confidence level), then $t_{0.025,17} = -2.11$. Comparing this to T_0, we see that $t_{0.025,17} < T_0$, which doesn't satisfy the criterion to reject the null hypothesis. In fact, for this assumption for α, the p-value is greater than α. Therefore, if we want to have at least a 95% confidence level, then we conclude that there is insufficient evidence to reject the null hypothesis.

The verification of these results is obtained with interactive graphic IG2–4, which yields Table 2.8.

Example 2.18 Case 6 of Table 2.3

From a population, 17 samples are randomly selected. The population is assumed to be normally distributed. The variance of the sample is $s^2 = 0.014$. The requirement is

Table 2.8 Hypothesis test results using interactive graphic IG2–4 for Example 2.17

$T_0 = -1.959$ H_0: $\mu_1 - \mu_2 = \delta_0 = 0.0$	Rejection criteria		
Confidence level	H_A: $\mu_1 - \mu_2 \neq \delta_0$	H_A: $\mu_1 - \mu_2 > \delta_0$	H_A: $\mu_1 - \mu_2 < \delta_0$
90% ($\alpha = 0.1$)	$T_0 < -1.74$ or $T_0 > 1.74$	$T_0 > 1.333$	$T_0 < -1.333$
95% ($\alpha = 0.05$)	$T_0 < -2.11$ or $T_0 > 2.11$	$T_0 > 1.74$	$T_0 < -1.74$
99% ($\alpha = 0.01$)	$T_0 < -2.898$ or $T_0 > 2.898$	$T_0 > 2.567$	$T_0 < -2.567$
99.9% ($\alpha = 0.001$)	$T_0 < -3.965$ or $T_0 > 3.965$	$T_0 > 3.646$	$T_0 < -3.646$
p-value	0.06676	0.03338	0.03338

that the population's variance cannot exceed 0.01. Based on this sample, we would like to determine if the requirement that $\sigma^2 = 0.01$ is being met at the 95% confidence level.

The null hypothesis statement is H_0: $\sigma^2 = 0.01$ with the alternative hypothesis being H_A: $\sigma^2 > 0.01$. Thus, $\sigma_0^2 = 0.01$ and the test statistic is

$$X_0 = \frac{(n-1)s^2}{\sigma_0^2} = \frac{(17-1)\times 0.014}{0.01} = 22.4$$

From Table 2.3, the rejection criterion is $X_0 > \chi_{0.05,16}$. For $\alpha = 0.05$, we find from interactive graphic IG2–2 that $\chi_{0.05,16} = 26.3$, and, therefore, this criterion is not satisfied, and the null hypothesis is not rejected. The p-value is determined from Eq. (2.119) as

$$p - \text{value} = \Psi_\chi(22.4, 16) = 0.1307$$

where we have used interactive graphic IG2–2 to obtain Ψ_χ. Thus, the p-value $> \alpha$ confirming that there is insufficient evidence that the null hypothesis should be rejected.

The verification of these results is obtained with interactive graphic IG2–4, which yields Table 2.9.

Example 2.19 Case 7 of Table 2.3

The variability of two independent processes is being investigated. One process has a standard deviation $s_1 = 1.81$ and the second a standard deviation $s_2 = 2.07$. Sixteen samples were used to determine each standard deviation and we assume that the data used to determine them came from normally distributed populations. We would like to determine if there are any differences in the processes based on these standard deviations.

The null hypothesis statement is H_0: $\sigma_1^2 = \sigma_2^2$ with the alternative hypothesis being H_A: $\sigma_1^2 \neq \sigma_2^2$. The test statistic is

Table 2.9 Hypothesis test results using interactive graphic IG2–4 for Example 2.18

$X_0 = 22.4$ $H_0 : \sigma^2 = \sigma_0^2 = 0.01$	Rejection criteria		
Confidence level	$H_A : \sigma^2 \neq \sigma_0^2$	$H_A : \sigma^2 > \sigma_0^2$	$H_A : \sigma^2 < \sigma_0^2$
90% ($\alpha = 0.1$)	$X_0 < 7.962$ or $X_0 > 26.3$	$X_0 > 23.54$	$X_0 < 9.312$
95% ($\alpha = 0.05$)	$X_0 < 6.908$ or $X_0 > 28.85$	$X_0 > 26.3$	$X_0 < 7.962$
99% ($\alpha = 0.01$)	$X_0 < 5.142$ or $X_0 > 34.27$	$X_0 > 32$	$X_0 < 5.812$
99.9% ($\alpha = 0.001$)	$X_0 < 3.636$ or $X_0 > 41.31$	$X_0 > 39.25$	$X_0 < 3.942$
p-value	0.2614	0.1307	0.1307

Table 2.10 Hypothesis test results using interactive graphic IG2–4 for Example 2.19

$F_0 = 0.7646$ $H_0: \sigma_1^2 = \sigma_2^2$	Rejection criteria		
Confidence level	$H_A: \sigma_1^2 \neq \sigma_2^2$	$H_A: \sigma_1^2 > \sigma_2^2$	$H_A: \sigma_1^2 < \sigma_2^2$
90% ($\alpha = 0.1$)	$F_0 < 0.4161$ or $F_0 > 2.403$	$F_0 > 1.972$	$F_0 < 0.507$
95% ($\alpha = 0.05$)	$F_0 < 0.3494$ or $F_0 > 2.862$	$F_0 > 2.403$	$F_0 < 0.4161$
99% ($\alpha = 0.01$)	$F_0 < 0.2457$ or $F_0 > 4.07$	$F_0 > 3.522$	$F_0 < 0.2839$
99.9% ($\alpha = 0.001$)	$F_0 < 0.1597$ or $F_0 > 6.264$	$F_0 > 5.535$	$F_0 < 0.1807$
p-value	0.6098	0.3049	0.3049

$$F_0 = \frac{s_1^2}{s_2^2} = \left(\frac{1.81}{2.07}\right)^2 = 0.7646$$

The p-value is obtained from Eq. (2.120) as

$$p - \text{value} = 2\Psi_f(0.8467, 15, 15) = 0.6098$$

where we have used interactive graphic IG2–2 to obtain Ψ_f. Since the p-value is considerably larger than 0.1, we do not reject the null hypothesis.

The verification for these results is obtained with interactive graphic IG2–4, which yields Table 2.10.

2.9.4 *Type I and Type II Errors*

The type I and type II errors are now determined. The type I error—reject H_0 when it is true—is simply α, the level of significance used in our confidence interval determination. This means that 100α % of the mean values obtained from different sets of data containing n samples would lead to the rejection of the null hypothesis H_0: $\mu = \mu_0$ when the true mean is μ_0. Therefore, the p-value is also the risk of incorrectly rejecting H_0.

The probability of a type II error, denoted β, can be determined only if we have a specific alternative hypothesis. Suppose that for a specified α and n the null hypothesis is H_0: $\mu = \mu_0$. To determine β, we assume that we have obtained the mean $\bar{x}$ from n samples from a population that is $N(\mu_1, \sigma^2/n)$. Then, we define β as

$$\beta = P\left(-z_{\alpha/2} \leq \frac{\bar{x} - \mu_0}{\sigma/\sqrt{n}} \leq z_{\alpha/2}\right) \tag{2.121}$$

We define the normal standard variable

$$Z = \frac{\bar{x} - \mu_1}{\sigma/\sqrt{n}}$$

Noting that

$$\frac{\bar{x} - \mu_0}{\sigma/\sqrt{n}} = \frac{\bar{x} - \mu_1 + \mu_1 - \mu_0}{\sigma/\sqrt{n}} = Z + z_d$$

where

$$z_d = \frac{\mu_1 - \mu_0}{\sigma/\sqrt{n}}$$

we find that Eq. (2.121) can be written as

$$\begin{aligned} \beta &= P\left(-z_{\alpha/2} \le Z + z_d \le z_{\alpha/2}\right) \\ &= P\left(-z_{\alpha/2} - z_d \le Z \le z_{\alpha/2} - z_d\right) \\ &= \Phi\left(z_{\alpha/2} - z_d\right) - \Phi\left(-z_{\alpha/2} - z_d\right) \end{aligned} \tag{2.122}$$

As stated previously, the quantity $1 - \beta$ is called the *power* of the test and is the probability of rejecting the null hypothesis H_0 when H_A is true. Thus, the smaller the magnitude of β the larger this probability.

These results are plotted in what are called operating characteristic (OC) curves as shown in Fig. 2.18. It is seen from this figure that for a given value of d, increasing the number of samples n decreases β. Also, for a given sample size, increasing d decreases β.

For numerical work, these curves can be accessed for any reasonable combination of n, α, and d in interactive graphic IG2–5. This interactive graphic also computes for a given α and d the number of samples required to attain a desired β.

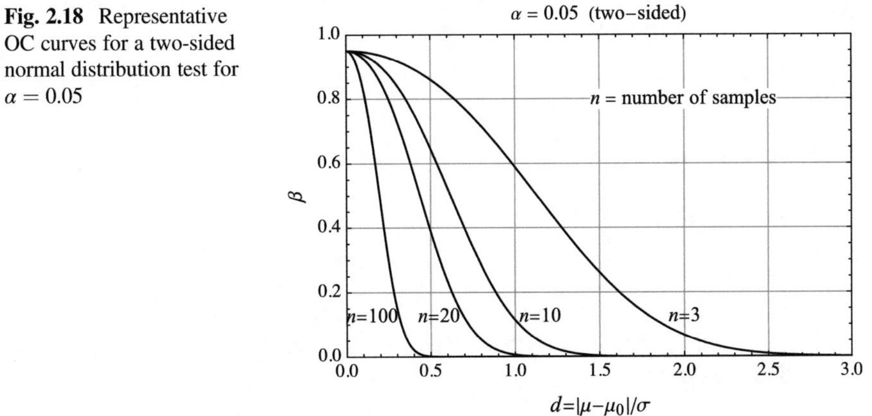

Fig. 2.18 Representative OC curves for a two-sided normal distribution test for $\alpha = 0.05$

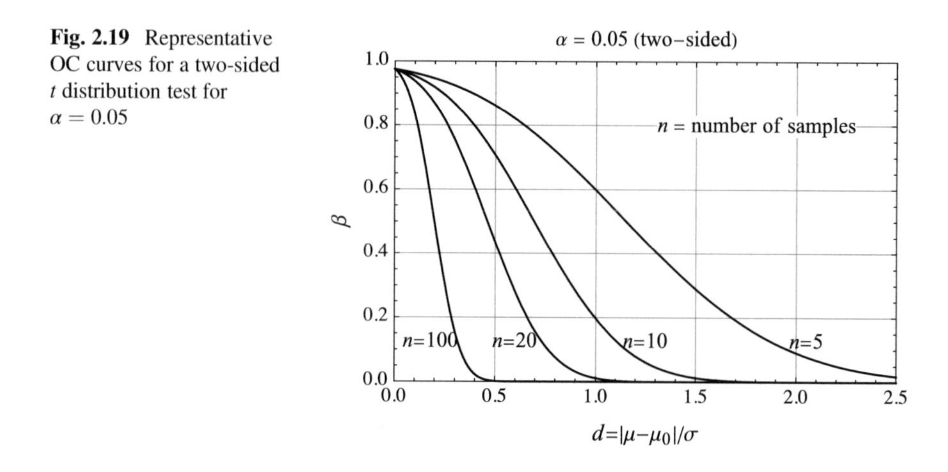

Fig. 2.19 Representative OC curves for a two-sided t distribution test for $\alpha = 0.05$

For other probability distributions the generation of OC curves is a little more complex. We have included the t distribution, which is shown in Fig. 2.19. For numerical work, these curves also can be accessed for any reasonable combination of n, α, and d in interactive graphic IG2–5. Since, for the t distribution, σ is not known, one may have to guess, use prior knowledge, or use the value of s from the samples to estimate this value.

How one typically uses the OC curves is shown in the following example.

Example 2.20

It is known that the standard deviation σ of a process is 2. A null hypothesis H_0: $\mu = 50$ was posed for $\alpha = 0.05$ and for a sample size $n = 25$. We are interested in determining the type II error if, indeed, the true mean is $\mu = 51$. Then, using interactive graphic IG2–5 for the two-sided normal distribution, with $d = |51–50|/2 = 0.5$, $\alpha = 0.05$, and $n = 25$ it is found that $\beta = 0.295$. Thus, if the true mean is 51, there is approximately a 30% chance that this will not be detected with 25 samples.

If one wants to reduce this type II error to, say, $\beta = 0.1$, we again use interactive graphic IG2–5 with $d = 0.5$, $\alpha = 0.05$, and $\beta = 0.1$ and find that the type II error decreases to 10% with a sample size of $n = 43$.

2.10 Exercises

Sect. 2.3.2

2.1. A test on 48 samples yields a mean value of 41,600 and a standard deviation of 16,200. If it can be assumed that the measured standard deviation is a very good estimate of σ, then:

(a) What is the confidence interval on the mean at the 95% level?
(b) What is the upper confidence limit at the 98% level.

2.2. The variation about a mean for a process that has a known variance equal to 0.1 is ±0.25 at the 99% confidence level.

(a) How many samples were used to get this variation about the mean?
(b) Determine the sample size if the variation about the mean is to be reduced to 0.15 at the 99% confidence level.

Sect. 2.3.5

2.3. It is found that the lifetime (in days) of a component is lognormally distributed with a mean of 1.5 and a standard variation of 0.25.

(a) Determine the expected value of the component lifetime.
(b) Determine the standard deviation of the lifetimes.
(c) Determine the probability that the lifetime will exceed 3 days.

Sect. 2.4

2.4. A manufacturing process requires that the weight of its product must be less than a certain value. An analysis of 22 random samples of this product found that the standard deviation is 0.0143. If it can be assumed that the samples are normally distributed, what is the upper limit on the standard deviation at the 95% confidence level.

Sect. 2.5

2.5. It is found from a series of 45 tests on the weights of an artifact that the mean value is 1.972 and the standard deviation is 0.061.

(a) Find the confidence interval of the mean at the 95% level.
(b) Find the confidence interval of the mean at the 99% level.
(c) Do you consider the differences between the confidence intervals in (a) and (b) meaningful?

2.6. From a strength test on 16 samples, it is found that the mean is 568 and the standard deviation is 141. What are the confidence limits at the 99% level?

Sect. 2.6

2.7. A test on a characteristic of two different products revealed the following. From 78 samples of product 1 the mean value was 4.25 and its known standard deviation was 1.30, and from 88 samples of product 2 the mean value was 7.14 and its known standard deviation was 1.68. Determine if product 2 has the larger characteristic at the 95% confidence level.

2.8. Two sets of data taken from a normal population reveal the following. From set 1, which has 24 samples, the mean is 9.005 and the standard deviation is 3.438 and from set 2, which has 34 samples, the mean is 11.864 and the standard deviation is 3.305. Determine the confidence limits on the differences in the mean values using the assumptions about the variances given by cases 3, 4, and 5 in Table 2.2 to determine if there are any differences in their means at the 99% confidence level.

2.9. A paired test is run with 20 observations. The mean is −0.87 and the standard deviation is 2.9773. Determine the confidence interval at the 95% level. What conclusion can be inferred from this interval?

Sect. 2.7

2.10. To determine the presence of a chemical in a river, tests were run at two different locations. In the first location 15 samples were collected and the standard deviation was found to be 3.07. At the second location, from 12 samples the standard deviation was found to be 0.80. Determine the confidence interval at the 98% confidence level and from this interval determine if the standard deviations are different.

Sect. 2.9.3

2.11. It was found from six samples that the mean value was 6.68 and the standard deviation was 0.20. From these measurements, can one conclude that the mean value is less than 7.0?

2.12. The mean values of the same characteristic from two processes are compared to determine if there is a difference between them. The first process had a mean value of 0.37 and a standard deviation of 0.25, which were obtained from 544 samples. The second process had a mean value of 0.40 and a standard deviation of 0.26, which were obtained from 581 samples. The difference between the mean values of process one and process two is −0.015. Is this difference due to the mean of process two being greater than that of process one? Since the sample size is so large, it is reasonable to assume that the measured standard deviations are a good representation of their respective population means.

2.13. The mean values of the same characteristic from two processes are compared to determine if there is a difference between them. The first process had a mean value of 44.1 and a standard deviation of 10.09, which were obtained from 10 samples. The second process had a mean value of 32.3 and a standard deviation of 8.56, which were also obtained from 10 samples. Nothing is known about the respective population standard deviations. Determine if process one has a greater mean than process two.

2.14. The standard deviation of a process must not exceed 0.15. Nineteen samples are measured, and it is found that the variance is 0.025. Are we justified in saying that process is not exceeding this limit with 95% confidence?

2.15. A manufacturing process is known to have a standard deviation of 2.95. Modifications are made to the process to reduce the standard deviation.

From a sample of 28 measurements, it is found that the standard deviation is 2.65. Determine if at the 95% confidence level the process has been improved.

2.16. A manufacturing company tests the strength of two different mixtures. One mixture is tested with 21 samples and the tests result in a standard deviation of 41.3. The second mixture is tested with 16 samples and results in a standard deviation of 26.2. If it can be assumed that the strengths have a normal distribution, determine at the 1% level of significance if the first mixture has more variability than the second mixture.

2.17. An experiment is conducted between two groups. It was found from 4 samples that the first group had a standard deviation of 39.0. In the second group 28 samples were used, and it was found that the standard deviation was 15.264. Is there evidence that the variance of the second group is greater than that for the first group? Assume that the respective distributions are normal.

Sect. 2.9.4 [These exercises most easily done with IG2–5]

2.18. The standard deviation of a process is known to be 5. A null hypothesis H_0: $\mu = 70$ and an alternative H_A: $\mu > 70$ are posed for $\alpha = 0.05$ and for a sample size $n = 30$.

   (a) Determine the percentage chance that if the true mean is 71 this will not be detected with 30 samples.
   (b) If one wants to increase the power of the test to 0.9, what must the sample size be?

2.19. A manufacturer has a requirement that a characteristic of a certain item is to have a mean of 125 and a standard deviation of 20. A supplier says that this item can be provided having the same standard deviation but an average value of 135. If the supplier wants a $\beta = 0.2$ and the manufacturer wants an $\alpha = 0.05$, what sample size must be used by the supplier to obtain this average to satisfy these requirements. Assume a two-sided distribution.

2.20. A product is to operate at a value that is less than or equal to 1.5. The standard deviation of the system is assumed to be 0.10. Ten samples are selected, and the tester would like to know:

   (a) How likely is it that a value of 1.6 will not be detected when $\alpha = 0.05$. Use the t test.
   (b) How many samples are needed to have $\beta = 0.05$.

Reference

Satterthwaite FE (1946) An approximate distribution of estimates of variance components. Biom Bull 2:110–114

Chapter 3
Regression Analysis and the Analysis of Variance

In this chapter, we provide derivations of the formulas for simple and multiple linear regression. In obtaining these results, the partitioning of the data using the sum of squares identities and the analysis of variance (ANOVA) are introduced. The confidence intervals of the model's parameters are determined and how an analysis of the residuals is used to confirm that the model is appropriate. Prior to obtaining the analytical results, we discuss why it is necessary to plot data before attempting to model it, state general guidelines and limits of a straight-line model to data, and show how one determines whether plotted data that appear to be nonlinear could be intrinsically linear. Hypothesis tests are introduced to determine which parameters in the model are the most influential.

3.1 Introduction

Regression analysis uses a model to determine the best relationship between one or more independent variables and the dependent variable, often called the response variable. The independent variable is called the *regressor*. When there is one independent variable the analysis is called *simple regression* and when there is more than one independent variable, it is called *multiple regression*. The model will contain unknown parameters that are determined from the data. *Linear regression* uses a model that is composed of a linear combination of the unknown parameters regardless of how the independent variables enter the model. In many situations, regression analysis is used to examine data wherein the underlying nature is not known or well understood. For this reason, the resulting regression model is often called an *empirical model*, an approximation to something unknown or complex.

Supplementary Information The online version contains supplementary material available at [https://doi.org/10.1007/978-3-031-05010-7_3].

E. B. Magrab, *Engineering Statistics*, https://doi.org/10.1007/978-3-031-05010-7_3

When the relation between the independent variables and the dependent variable has a random component to them, the relation will not be deterministic (exact), and the model must include the random component and its statistical properties; that is, the random component must, itself, be modeled.

Before proceeding with the formal introduction to simple linear regression, it is worth mentioning some restrictions, good practices, clarifications, and limitations.

1. Care should be taken in selecting variables with which to construct regression relations so that one develops a model for which the variables are causally related.

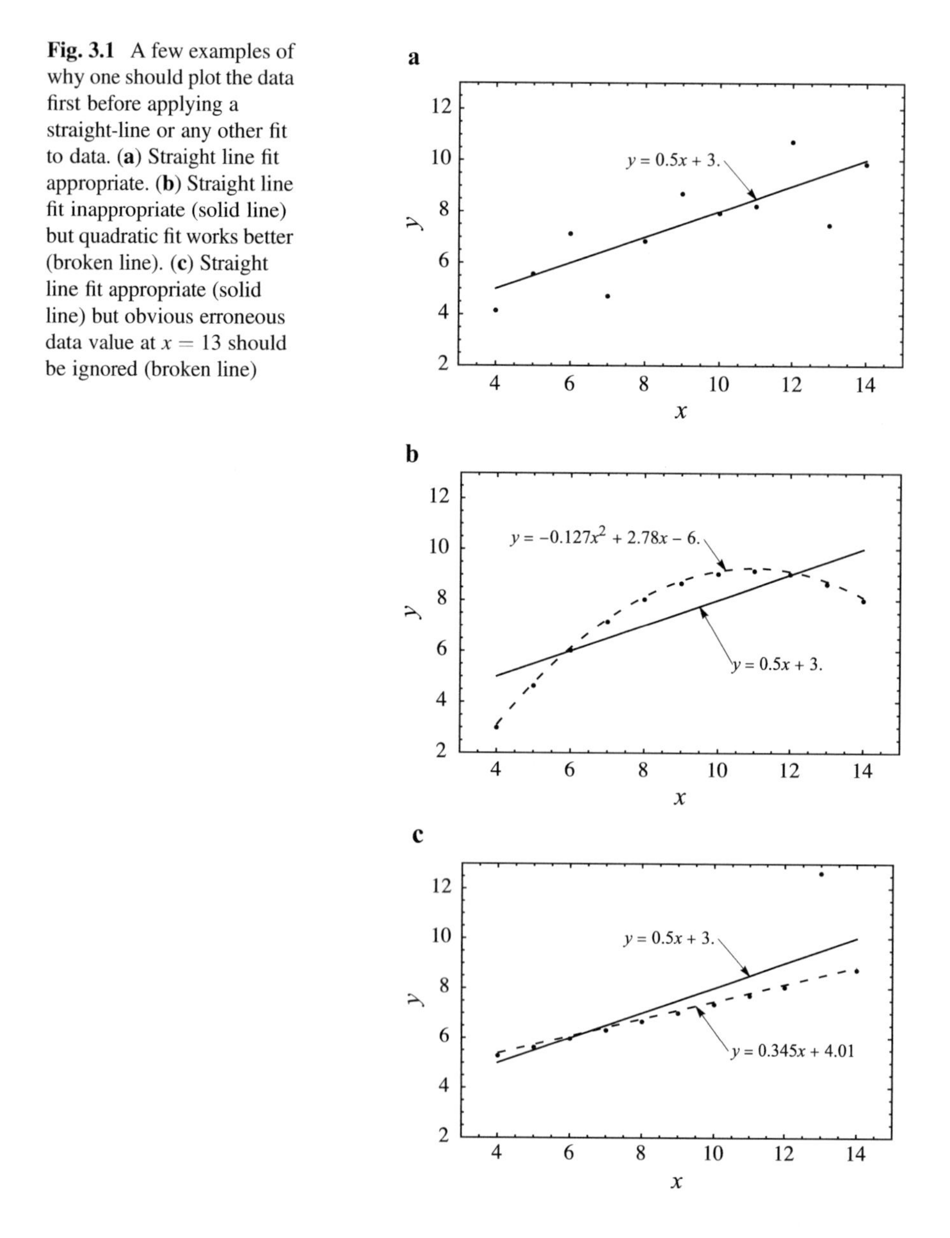

Fig. 3.1 A few examples of why one should plot the data first before applying a straight-line or any other fit to data. (**a**) Straight line fit appropriate. (**b**) Straight line fit inappropriate (solid line) but quadratic fit works better (broken line). (**c**) Straight line fit appropriate (solid line) but obvious erroneous data value at $x = 13$ should be ignored (broken line)

For example, automobile gas mileage versus automobile's weight seems to make sense as opposed to automobile gas mileage versus automobile color, which doesn't. A modeled association between variables that fit the data well does not necessarily mean that a causal relationship exists between them.
2. The regression relations are valid only for the range of the data used to construct them: avoid extrapolation; that is, using the model outside this range.
3. One should always plot the data before attempting any curve-fitting procedure because while it is always possible to fit a straight line to a set of data it may not be appropriate to do so. Consider the examples shown in Fig. 3.1. If we assume that each set of data was fit with a straight line prior to plotting them, it is seen that a straight-line assumption gives the same set of parameters for their respective fitted lines. Clearly, for two of these sets, this is inappropriate.
4. As mentioned previously, linear regression uses a model that is composed of a linear combination of the parameters appearing in the model. Thus, for example, if b_0, b_1, and b_2 are the unknown parameters, then

$$\begin{aligned} y &= b_0 + b_1 \ln x \\ y &= b_0 + b_1 \sin x + b_2 x^2 \end{aligned} \tag{3.1}$$

are linear regression models because the unknown parameters b_j are linear but

$$\begin{aligned} y &= x^{b_0} \cos(b_1 x) \\ y &= x^{b_0} - (b_1)^{-x} \end{aligned} \tag{3.2}$$

are not linear models. Additional examples are given in Sect. 3.3.

3.2 Simple Linear Regression and the Analysis of Variance

3.2.1 Simple Linear Regression

We assume that there is an input to a component, system, or process that produces a response. The input is considered an independent *deterministic* variable denoted x and the output, the response variable, is considered the dependent variable denoted y. For simple linear regression, the following linear relationship between the response variable and the input variable is assumed

$$y = \beta_0 + \beta_1 x \tag{3.3}$$

where the unknown parameters β_0 and β_1 are determined from measurements of the response. The parameter β_0 is the intercept and the parameter β_1 is the slope of the line and these parameters are called the *regression coefficients*. If $\beta_1 = 0$, the line is horizontal indicating that the independent variable x does not affect the dependent variable y; that is, they are unrelated.

In practice, the output measurements are subject to a certain amount of randomness. To account for this, the model given by Eq. (3.3) is modified as

$$y = \beta_0 + \beta_1 x + \varepsilon \tag{3.4}$$

where ε is an additive random variable that is assumed to be normally distributed as $\varepsilon \sim N(0,\sigma^2)$; that is, $E(\varepsilon) = 0$ and $\text{Var}(\varepsilon) = \sigma^2$, which is called the *error variance*. The quantity ε is a conceptual construct and not an observed one. Thus, for a deterministic input x, ε represents the randomness in the measured value y. Because of the introduction of ε, y is a random variable with distribution $N(\beta_0 + \beta_1 x,\sigma^2)$ since we have assumed that $E(\varepsilon) = 0$. Additionally, since $E(\varepsilon) = 0$ the variation of y will be distributed around the regression line.

We shall now introduce a method from which we can obtain estimates for β_0, β_1, and, eventually, σ^2. Consider a set of data S that is composed of the pairs $S = \{(x_1, y_1), (x_2,y_2), \ldots,(x_n,y_n)\}$. For each pair of data values,

$$y_i = \beta_0 + \beta_1 x_i + \varepsilon_i \quad i = 1, 2, .., n \tag{3.5}$$

where it is assumed that $\varepsilon_i \sim N(0,\sigma^2)$; that is, $E(\varepsilon_i) = 0$ and $\text{Var}(\varepsilon_i) = \sigma^2$. Equation (3.5) represents a system of n equations with two unknowns β_0 and β_1. Then, for $n > 2$ we have an overdetermined system of equations. One way that estimates for β_0 and β_1 can be obtained is to use the method of least squares, which finds the values of β_0 and β_1 that minimize the sum of the square of the residuals; that is,

$$G = \sum_{i=1}^{n} \varepsilon_i^2 = \sum_{i=1}^{n} (y_i - (\beta_0 + \beta_1 x_i))^2 \tag{3.6}$$

Thus, the least squares procedure will find the estimates of β_0 and β_1 that produce a line that minimizes the sum of squares of the vertical deviations from the points y_i to the line. The resulting parameter estimates are denoted $\widehat{\beta}_0$ and $\widehat{\beta}_1$. The expression in Eq. (3.6) is a function of two variables, which are determined from the minimization of G. From multivariable calculus, the minimization is obtained from the following operations

$$\frac{\partial G}{\partial \beta_o} = -2\sum_{i=1}^{n} (y_i - (\beta_0 + \beta_1 x_i)) = 0$$

$$\frac{\partial G}{\partial \beta_1} = -2\sum_{i=1}^{n} x_i(y_i - (\beta_0 + \beta_1 x_i)) = 0$$

which result in

$$\begin{aligned} \sum_{i=1}^{n} y_i - n\widehat{\beta}_0 - \widehat{\beta}_1 \sum_{i=1}^{n} x_i = 0 \\ \sum_{i=1}^{n} x_i y_i - \widehat{\beta}_0 \sum_{i=1}^{n} x_i - \widehat{\beta}_1 \sum_{i=1}^{n} x_i^2 = 0 \end{aligned} \tag{3.7}$$

The equations in Eq. (3.7) are known as the *normal equations*.

We introduce the following quantities[1] called the *sum of squares* (recall Eq. (1.13))

$$\begin{aligned} S_{xy} &= \sum_{i=1}^{n}(x_i-\overline{x})(y_i-\overline{y}) = \sum_{i=1}^{n} y_i(x_i-\overline{x}) = \sum_{i=1}^{n} x_i y_i - n\overline{xy} \\ S_{xx} &= \sum_{i=1}^{n}(x_i-\overline{x})^2 = \sum_{i=1}^{n} x_i^2 - n\overline{x}^2 \end{aligned} \tag{3.8}$$

where

$$\begin{aligned} \overline{y} &= \frac{1}{n}\sum_{i=1}^{n} y_i \\ \overline{x} &= \frac{1}{n}\sum_{i=1}^{n} x_i \end{aligned} \tag{3.9}$$

Then, Eq. (3.7) can be written as

$$\begin{aligned} \widehat{\beta}_0 + \widehat{\beta}_1\overline{x} &= \overline{y} \\ n\overline{x}\widehat{\beta}_0 + \widehat{\beta}_1\left(S_{xx}+n\overline{x}^2\right) &= S_{xy} + n\overline{xy} \end{aligned} \tag{3.10}$$

Solving Eq. (3.10) for $\widehat{\beta}_0$ and $\widehat{\beta}_1$, we find that

$$\begin{aligned} \widehat{\beta}_1 &= \frac{S_{xy}}{S_{xx}} \\ \widehat{\beta}_0 &= \overline{y} - \widehat{\beta}_1\overline{x} = \overline{y} - \frac{S_{xy}}{S_{xx}}\overline{x} \end{aligned} \tag{3.11}$$

The fitted regression line becomes

$$\widehat{y} = \widehat{\beta}_o + \widehat{\beta}_1 x = \overline{y} + \widehat{\beta}_1(x-\overline{x}) \tag{3.12}$$

where we have used the second equation of Eq. (3.11). Therefore, the original values y_i are now estimated by

---

[1] It is noted that

$$\sum_{i=1}^{n}(x_i-\overline{x})(y_i-\overline{y}) = \sum_{i=1}^{n} y_i(x_i-\overline{x}) - \overline{y}\sum_{i=1}^{n}(x_i-\overline{x}) = \sum_{i=1}^{n} y_i(x_i-\overline{x})$$

since $\sum_{i=1}^{n}(x_i-\overline{x}) = 0$.

Table 3.1 Intrinsically linear nature of several nonlinear equations where the straight-line parameters β_0 and β_1 are obtained from the fit to $y = \beta_0 + \beta_1 x$

Case	Original model	Transformed linear model	Parameters
1 Exponential	$q = \gamma_0 e^{\gamma_1 w}$	$\ln q = \ln \gamma_0 + \gamma_1 w$	$\gamma_0 = e^{\beta_0}$ $\gamma_1 = \beta_1$
2 Reciprocal	$q = \dfrac{1}{\gamma_0 + \gamma_1 w}$	$\dfrac{1}{q} = \gamma_0 + \gamma_1 w$	$\gamma_0 = \beta_0$ $\gamma_1 = \beta_1$
3 Power	$q = \gamma_0 w^{\gamma_1}$	$\ln q = \ln \gamma_0 + \gamma_1 \ln w$	$\gamma_0 = e^{\beta_0}$ $\gamma_1 = \beta_1$
4 Normal cdf	$q = \Phi\left(\dfrac{w - \mu}{\sigma}\right)$	$\Phi^{-1}(q) = -\dfrac{\mu}{\sigma} + \dfrac{1}{\sigma} w$	$\sigma = 1/\beta_1$ $\mu = -\beta_0/\beta_1$
5 Weibull cdf	$q = 1 - e^{-(w/\beta)^\gamma}$	$\ln \ln\left(\dfrac{1}{1-q}\right) = \gamma \ln \beta + \gamma \ln w$	$\gamma = \beta_1$ $\beta = e^{-\beta_0/\beta_1}$

$$\widehat{y}_i = \widehat{\beta}_0 + \widehat{\beta}_1 x_i \tag{3.13}$$

It is seen from Eq. (3.12) that the regression line always goes through the point $(\overline{x}, \overline{y})$; that is, it goes through the mean of the independent variable and the mean of the response variable.

Intrinsically Linear Data

Before proceeding, it is mentioned that Eq. (3.13) may be applicable to nonlinearly shaped trends in the data by recognizing that these data could be intrinsically linear after they have been transformed by a suitable change of variables. Consider the examples shown in Table 3.1 and Fig. 3.2. In Fig. 3.2, the data are represented as filled circles. These data are transformed using the relations shown in Table 3.1 and appear as filled circles in the inset figures. To these data in the inset figure, a straight line is fit and the straight-line parameters β_0 and β_1 are found. These parameters are then used to determine the unknown parameters in the original nonlinear equation and that equation is plotted as the solid line in the figure. Cases 4 and 5 show another way that one can find the parameters of a normal and Weibull distribution. If it appears that one these distributions is a reasonable choice, the data are placed in the form of a cumulative frequency distribution as shown in Fig. 1.3 and one proceeds from there. When these transformations are unsuccessful, one may be able to use a multi-regression analysis as discussed in Sect. 3.3 to fit the data with a polynomial or some other mathematical function. Figure 3.2 was obtained with interactive graphic IG3-1.

Continuing with the analysis of Eq. (3.13), the residual is defined as

$$e_i = y_i - \widehat{y}_i = y_i - \widehat{\beta}_0 - \widehat{\beta}_1 x_i \tag{3.14}$$

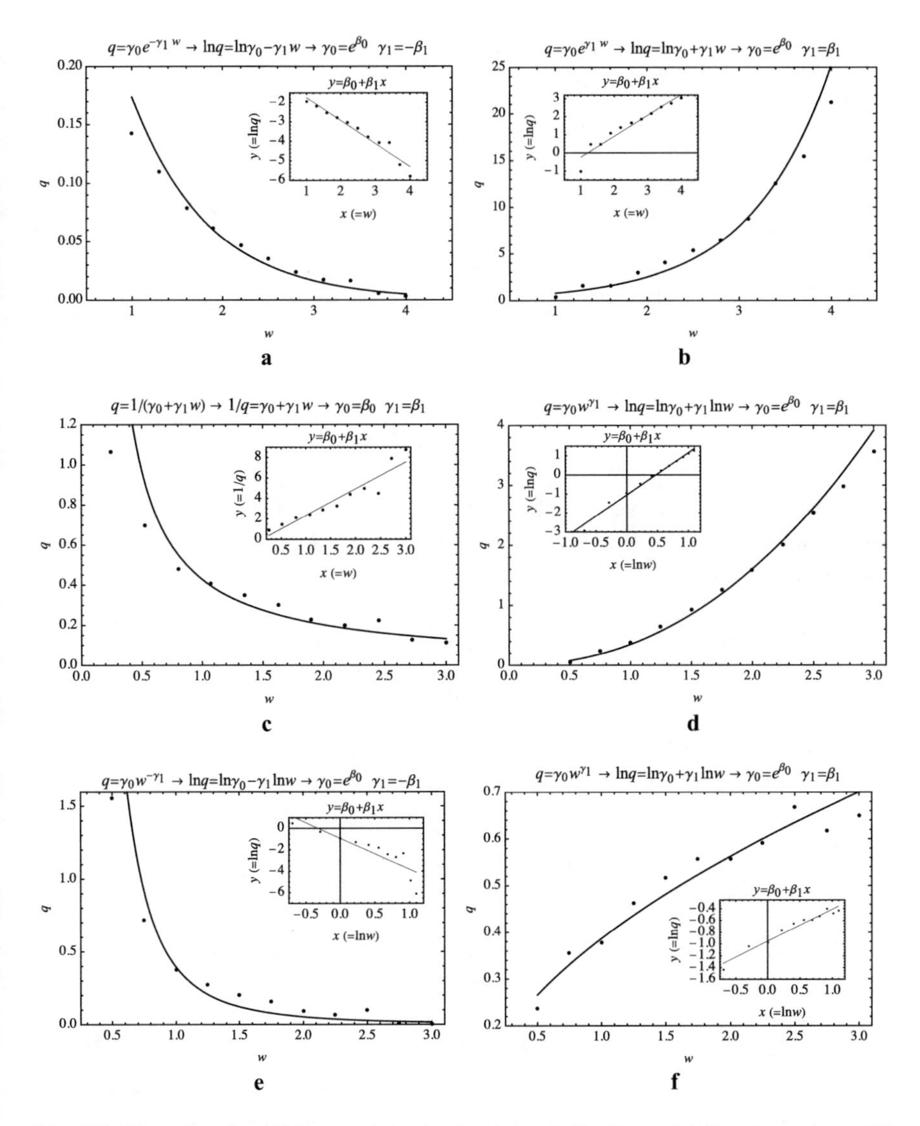

Fig. 3.2 Examples of nonlinear relations that are intrinsically linear. (**a**) Negative exponential function. (**b**) Positive exponential function. (**c**) Reciprocal-like function. (**d**) Power function with $\gamma_1 > 1$. (**e**) Power function with $\gamma_1 < 0$. (**f**) Power function with $0 < \gamma_1 < 1$. (**g**) Cumulative distribution function for the normal distribution. (**h**) Cumulative distribution function for the Weibull distribution

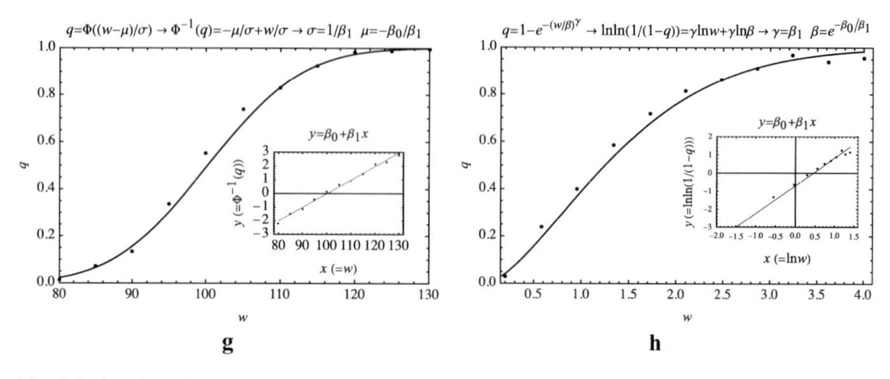

Fig. 3.2 (continued)

which is the difference between the observed value and the expected value. Using this definition, we shall note several of its properties. First, we find that the sum of the residuals is zero since

$$\sum_{i=1}^{n} e_i = \sum_{i=1}^{n} \left(y_i - \widehat{\beta}_0 - \widehat{\beta}_1 x_i\right) = \sum_{i=1}^{n} y_i - n\widehat{\beta}_0 - \widehat{\beta}_1 \sum_{i=1}^{n} x_i y_i = 0 \tag{3.15}$$

as indicated by the first equation of Eq. (3.7). In addition, the following sum of the product of the ith residual with the ith independent variable is zero; that is,

$$\sum_{i=1}^{n} x_i e_i = \sum_{i=1}^{n} x_i \left(y_i - \widehat{\beta}_0 - \widehat{\beta}_1 x_i\right) = \sum_{i=1}^{n} x_i y_i - \widehat{\beta}_0 \sum_{i=1}^{n} x_i - \widehat{\beta}_1 \sum_{i=1}^{n} x_i^2 = 0 \tag{3.16}$$

as indicated by the second equation of Eq. (3.7). It is also noted that

$$\sum_{i=1}^{n} \widehat{y}_i e_i = \sum_{i=1}^{n} \left(\widehat{\beta}_0 + \widehat{\beta}_1 x_i\right) e_i = \widehat{\beta}_0 \sum_{i=1}^{n} e_i + \widehat{\beta}_1 \sum_{i=1}^{n} x_i e_i = 0 \tag{3.17}$$

where we have used Eqs. (3.13), (3.15), and (3.16). Furthermore, we note that

$$\sum_{i=1}^{n} y_i = \sum_{i=1}^{n} \widehat{y}_i \tag{3.18}$$

since, from Eqs. (3.13) and (3.10)

$$\sum_{i=1}^{n}\widehat{y}_i = \sum_{i=1}^{n}\left(\widehat{\beta}_0 + \widehat{\beta}_1 x_i\right) = n\widehat{\beta}_0 + n\widehat{\beta}_1\overline{x}$$
$$= n\left(\overline{y} - \widehat{\beta}_1\overline{x}\right) + n\widehat{\beta}_1\overline{x} = n\overline{y} = \sum_{i=1}^{n} y_i$$

We shall now show that $\widehat{\beta}_1$ is an unbiased estimator of β_1 as defined by Eq. (1.36). Using Eqs. (3.8), (3.11), and (3.5), and the fact that x_i is deterministic, we have

$$\begin{aligned}
E\left(\widehat{\beta}_1\right) &= E\left(\frac{S_{xy}}{S_{xx}}\right) = E\left[\frac{1}{S_{xx}}\sum_{i=1}^{n} y_i(x_i - \overline{x})\right] \\
&= E\left[\frac{1}{S_{xx}}\sum_{i=1}^{n}(x_i - \overline{x})(\beta_0 + \beta_1 x_i + \varepsilon_i)\right] \\
&= \frac{1}{S_{xx}}\left[\beta_0\sum_{i=1}^{n}(x_i - \overline{x}) + \beta_1\sum_{i=1}^{n}x_i(x_i - \overline{x}) + \sum_{i=1}^{n}(x_i - \overline{x})E(\varepsilon_i)\right] \\
&= \frac{\beta_1}{S_{xx}}\left(\sum_{i=1}^{n}x_i^2 - \overline{x}\sum_{i=1}^{n}x_i\right) = \frac{\beta_1}{S_{xx}}\left(\sum_{i=1}^{n}x_i^2 - n\overline{x}^2\right) = \beta_1
\end{aligned} \tag{3.19}$$

since $E(\varepsilon_i) = 0$ and $\sum_{i=1}^{n}(x_i - \overline{x}) = 0$. Next, we show that $\widehat{\beta}_0$ is an unbiased estimator of β_0. Again, using Eqs. (3.11) and (3.5) and the fact that x_i is deterministic, we have

$$\begin{aligned}
E\left(\widehat{\beta}_o\right) &= E\left(\overline{y} - \widehat{\beta}_1\overline{x}\right) = \frac{1}{n}\sum_{i-1}^{n}E(y_i) - \overline{x}E\left(\widehat{\beta}_1\right) \\
&= \frac{1}{n}\sum_{i-1}^{n}E(\beta_0 + \beta_1 x_i + \varepsilon_i) - \overline{x}\beta_1 \\
&= \frac{1}{n}\sum_{i-1}^{n}(\beta_0 + \beta_1 x_i) - \overline{x}\beta_1 = \beta_0
\end{aligned} \tag{3.20}$$

Before determining the variance of $\widehat{\beta}_0$ and $\widehat{\beta}_1$, we note from Eq. (1.31) that

$$\begin{aligned}
\text{Var}(y_i) &= \text{Var}(\beta_0 + \beta_1 x_i + \varepsilon_i) \\
&= \text{Var}(\beta_0) + x_i^2\text{Var}(\beta_1) + \text{Var}(\varepsilon_i) \\
&= \sigma^2
\end{aligned}$$

since $\text{Var}(\beta_0) = \text{Var}(\beta_1) = 0$. We also note that

$$\text{Var}(\bar{y}) = \text{Var}\left(\frac{1}{n}\sum_{i=1}^{n} y_i\right) = \frac{1}{n^2}\sum_{i=1}^{n}\text{Var}(y_i) = \frac{1}{n^2}\sum_{i=1}^{n}\sigma^2 = \frac{\sigma^2}{n}$$

Then, using Eqs. (1.31) and (3.8), the variance of $\widehat{\beta}_1$ is

$$\begin{aligned}\text{Var}\left(\widehat{\beta}_1\right) &= \text{Var}\left(\frac{S_{xy}}{S_{xx}}\right) = \text{Var}\left[\sum_{i=1}^{n}\left(\frac{x_i - \bar{x}}{S_{xx}}\right)y_i\right] \\ &= \sum_{i=1}^{n}\left[\left(\frac{x_i - \bar{x}}{S_{xx}}\right)^2 \text{Var}\,(y_i)\right] = \frac{\sigma^2}{S_{xx}^2}\sum_{i=1}^{n}(x_i - \bar{x})^2 = \frac{\sigma^2}{S_{xx}}\end{aligned} \tag{3.21}$$

Using Eqs. (1.31), (3.11), and the previous results, the variance of $\widehat{\beta}_0$ is

$$\begin{aligned}\text{Var}\left(\widehat{\beta}_0\right) &= \text{Var}\left(\bar{y} - \widehat{\beta}_1\bar{x}\right) = \text{Var}(\bar{y}) + \bar{x}^2\text{Var}\left(\widehat{\beta}_1\right) \\ &= \frac{\sigma^2}{n} + \frac{\sigma^2\bar{x}^2}{S_{xx}} = \sigma^2\left(\frac{1}{n} + \frac{\bar{x}^2}{S_{xx}}\right)\end{aligned} \tag{3.22}$$

Therefore,

$$\begin{aligned}\widehat{\beta}_1 &\sim N\left(\beta_1, \sigma^2/S_{xx}\right) \\ \widehat{\beta}_0 &\sim N\left(\beta_0, \sigma^2\left(1/n + \bar{x}^2/S_{xx}\right)\right)\end{aligned} \tag{3.23}$$

We now use Eqs. (3.12) and (3.21) to obtain the variance of $\widehat{y}$ as follows

$$\begin{aligned}\text{Var}(\widehat{y}) &= \text{Var}\left(\bar{y} + \widehat{\beta}_1(x - \bar{x})\right) = \text{Var}(\bar{y}) + \text{Var}\left(\widehat{\beta}_1(x - \bar{x})\right) \\ &= \frac{\sigma^2}{n} + (x - \bar{x})^2\frac{\sigma^2}{S_{xx}} = \sigma^2\left(\frac{1}{n} + \frac{(x - \bar{x})^2}{S_{xx}}\right)\end{aligned} \tag{3.24}$$

which is a function of x. The variance of $\widehat{y}$ has a minimum at $x = \bar{x}$.

Partitioning of the Variability

Consider the following identity

$$(y_i - \bar{y}) = (\widehat{y}_i - \bar{y}) + (y_i - \widehat{y}_i) \tag{3.25}$$

where y_i is the measured response, $\bar{y}$ is the average value of the n measured responses, and $\widehat{y}_i$ is the predicted response. In Eq. (3.25), the terms have the following interpretations:

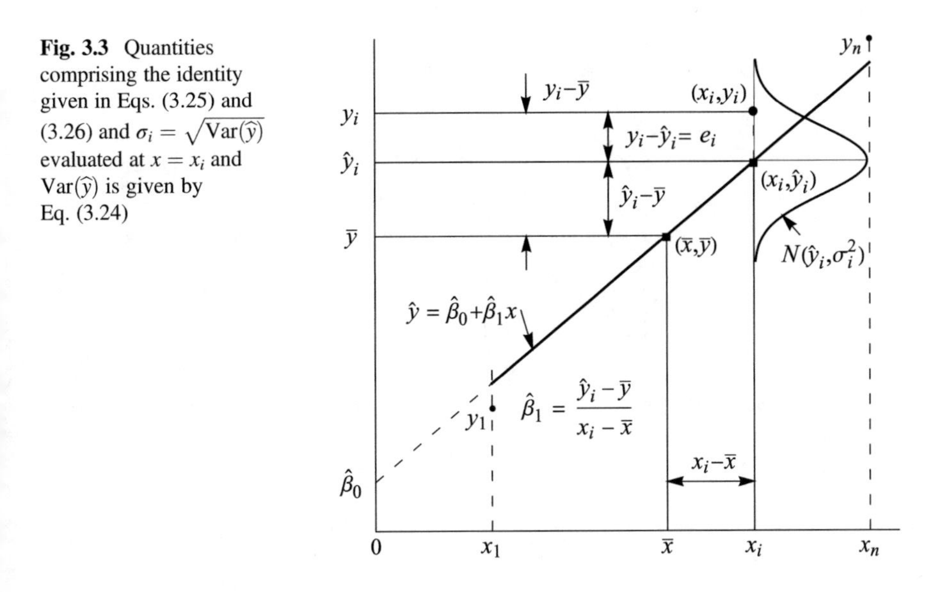

Fig. 3.3 Quantities comprising the identity given in Eqs. (3.25) and (3.26) and $\sigma_i = \sqrt{\mathrm{Var}(\hat{y})}$ evaluated at $x = x_i$ and $\mathrm{Var}(\hat{y})$ is given by Eq. (3.24)

$y_i - \bar{y}$	is the deviation of the measured response from its mean
$\hat{y}_i - \bar{y}$	is the deviation of the regression line value from its mean and is the explained deviation
$y_i - \hat{y}_i = e_i$	is the deviation of the regression line value from the measured value, which has been defined as the residual, and is the unexplained (random) deviation

These terms are shown in Fig. 3.3.

3.2.2 *Analysis of Variance (ANOVA)*

We now consider an expression called the *sum of squares identity*. The rationale for this identity is to determine whether all the variation in the data is attributable to random error (chance) or if some of the variation in the data is attributable to chance and some of it is attributable to the model's parameters. The use of these partitioned results to obtain an expression for an estimate of the variance is termed an **an**alysis **o**f **va**riance (ANOVA). Once we have an estimate of the variance, a hypothesis test can be performed on the model's parameters. The sum of squares identity is

$$\sum_{i=1}^{n}(y_i - \bar{y})^2 = \sum_{i=1}^{n}(\hat{y}_i - \bar{y})^2 + \sum_{i=1}^{n}(y_i - \hat{y}_i)^2 \tag{3.26}$$

To prove this identity, we substitute Eq. (3.25) into the left-hand portion of Eq. (3.26) and obtain

$$\begin{aligned}\sum_{i=1}^{n}(y_i-\bar{y})^2&=\sum_{i=1}^{n}((\widehat{y}_i-\bar{y})+(y_i-\widehat{y}_i))^2\\&=\sum_{i=1}^{n}\left((\widehat{y}_i-\bar{y})^2+(y_i-\widehat{y}_i)^2+2(\widehat{y}_i-\bar{y})(y_i-\widehat{y}_i)\right)\\&=\sum_{i=1}^{n}(\widehat{y}_i-\bar{y})^2+\sum_{i=1}^{n}(y_i-\widehat{y}_i)^2+2\sum_{i=1}^{n}\widehat{y}_ie_i-2\bar{y}\sum_{i=1}^{n}e_i\\&=\sum_{i=1}^{n}(\widehat{y}_i-\bar{y})^2+\sum_{i=1}^{n}(y_i-\widehat{y}_i)^2\end{aligned}$$

where we have used Eqs. (3.15) and (3.17). We identify the terms in Eq. (3.26) as follows

$$\begin{aligned}SS_T&=\sum_{i=1}^{n}(y_i-\bar{y})^2=\sum_{i=1}^{n}y_i^2-n\bar{y}^2\ \text{(total sum of squares)}\\SS_R&=\sum_{i=1}^{n}(\widehat{y}_i-\bar{y})^2\ \text{(regression sum of squares)}\\SS_E&=\sum_{i=1}^{n}(y_i-\widehat{y}_i)^2\ \text{(error sum of squares)}\end{aligned}\tag{3.27}$$

Then Eq. (3.26) can be written as

$$SS_T=SS_R+SS_E\tag{3.28}$$

To find an estimate of the error variance σ^2, we determine the expected value of the terms in Eq. (3.28); that is,

$$E(SS_E)=E(SS_T)-E(SS_R)\tag{3.29}$$

We first note that SS_R can be written as

$$\begin{aligned}SS_R&=\sum_{i=1}^{n}(\widehat{y}_i-\bar{y})^2=\sum_{i=1}^{n}\left(\bar{y}+\widehat{\beta}_1(x_i-\bar{x})-\bar{y}\right)^2\\&=\widehat{\beta}_1^2\sum_{i=1}^{n}(x_i-\bar{x})^2=\widehat{\beta}_1^2S_{xx}=\widehat{\beta}_1S_{xy}\end{aligned}\tag{3.30}$$

where we have used Eqs. (3.8), (3.11), and (3.12). Then, using Eqs. (1.15), (3.19), and (3.21)

$$
\begin{aligned}
E(SS_R) &= E\left(\widehat{\beta}_1^2 S_{xx}\right) = S_{xx} E\left(\widehat{\beta}_1^2\right) = S_{xx}\left(\mathrm{Var}\left(\widehat{\beta}_1\right) + \left(E\left(\widehat{\beta}_1\right)\right)^2\right) \\
&= S_{xx}\left(\frac{\sigma^2}{S_{xx}} + \beta_1^2\right) = \sigma^2 + S_{xx}\beta_1^2
\end{aligned}
\tag{3.31}
$$

Next, we determine the expected value of SS_T. Upon using Eqs. (1.15), (3.5), (3.10), and the fact that $E(\varepsilon_i) = 0$, we obtain

$$
\begin{aligned}
E(SS_T) &= E\left[\sum_{i=1}^{n} (y_i - \overline{y})^2\right] = E\left[\sum_{i=1}^{n} y_i^2 - n\overline{y}^2\right] = \sum_{i=1}^{n} E\left(y_i^2\right) - nE\left(\overline{y}^2\right) \\
&= \sum_{i=1}^{n} \left[\mathrm{Var}(y_i) + (E(y_i))^2\right] - n\left[\mathrm{Var}(\overline{y}) + (E(\overline{y}))^2\right] \\
&= \sum_{i=1}^{n} \left[\sigma^2 + (E(\beta_0 + \beta_1 x_i + \varepsilon_i))^2\right] - n\left[\frac{\sigma^2}{n} + \left(E\left(\widehat{\beta}_0 + \widehat{\beta}_1 \overline{x}\right)\right)^2\right] \\
&= \sum_{i=1}^{n} \left[\sigma^2 + (\beta_0 + \beta_1 x_i)^2\right] - \sigma^2 - n(\beta_0 + \beta_1 \overline{x})^2
\end{aligned}
$$

Carrying out the remaining manipulations, we find that

$$
E(SS_T) = (n-1)\sigma^2 + \beta_1^2 S_{xx}
\tag{3.32}
$$

Substituting Eqs. (3.31) and (3.32) into Eq. (3.29), we obtain

$$
\begin{aligned}
E(SS_E) &= E(SS_T) - E(SS_R) \\
&= (n-1)\sigma^2 + \beta_1^2 S_{xx} - \left(\sigma^2 + \beta_1^2 S_{xx}\right) \\
&= (n-2)\sigma^2
\end{aligned}
\tag{3.33}
$$

Thus, the variance σ^2 is

$$
\sigma^2 = E\left(\frac{SS_E}{n-2}\right)
\tag{3.34}
$$

and, therefore, an unbiased estimator of the error variance is

$$
\widehat{\sigma}^2 = \frac{SS_E}{n-2}
\tag{3.35}
$$

which has $n - 2$ degrees of freedom. Recalling Eq. (2.62) and the discussion preceding it, we see that

$$\frac{(n-2)\widehat{\sigma}^2}{\sigma^2} \sim \chi^2_{n-2} \tag{3.36}$$

has a chi square distribution with $n - 2$ degrees of freedom. We note that an unbiased estimate of the variance of y_i is

$$s_y^2 = \frac{1}{n-1}\sum_{i=1}^{n}(y_i - \overline{y})^2 = \frac{SS_T}{n-1}$$

and that

$$\frac{(n-1)s_y^2}{\sigma^2} \sim \chi^2_{n-1}$$

has a chi square distribution with $n - 1$ degrees of freedom. Therefore, from Eqs. (3.28) and (2.51), we see that SS_R has 1 degree of freedom.

A quantity called the *mean square* (MS) is used in linear regression and in the analysis of variance. It is defined as the sum of squares divided by the number of degrees of freedom. Then, the regression mean square is

$$\text{MSR} = \frac{SS_R}{1} = \sum_{i=1}^{n}(\widehat{y}_i - \overline{y})^2 \tag{3.37}$$

The error mean square is

$$\text{MSE} = \frac{SS_E}{n-2} = \widehat{\sigma}^2 = \frac{1}{n-2}\sum_{i=1}^{n}e_i^2 = \frac{1}{n-2}\sum_{i=1}^{n}(y_i - \widehat{y}_i)^2 \tag{3.38}$$

and the total mean square is

$$\text{MST} = \frac{SS_T}{n-1} = \frac{1}{n-1}\sum_{i=1}^{n}(y_i - \overline{y})^2 \tag{3.39}$$

Thus, from Eq. (3.31), the expected value of MSR, which has one degree of freedom, is

$$E(\text{MSR}) = E(SS_R) = \sigma^2 + S_{xx}\beta_1^2 \tag{3.40}$$

and from Eq. (3.34) the expected value of MSE, which has $n - 2$ degrees of freedom, is

$$E(\text{MSE}) = E\left(\frac{SS_E}{n-2}\right) = \sigma^2 \tag{3.41}$$

In the case where σ^2 is unknown, the variances for $\widehat{\beta}_0$ and $\widehat{\beta}_1$, respectively, can be estimated as

$$\begin{aligned} \text{Var}\left(\widehat{\beta}_0\right) &= \widehat{\sigma}^2\left(\frac{1}{n}+\frac{\overline{x}^2}{S_{xx}}\right) = \text{MSE}\left(\frac{1}{n}+\frac{\overline{x}^2}{S_{xx}}\right) \\ \text{Var}\left(\widehat{\beta}_1\right) &= \frac{\widehat{\sigma}^2}{S_{xx}} = \frac{\text{MSE}}{S_{xx}} \end{aligned} \tag{3.42}$$

and response variance can be estimated as

$$\text{Var}(\widehat{y}) = \widehat{\sigma}^2\left(\frac{1}{n}+\frac{(x-\overline{x})^2}{S_{xx}}\right) = \text{MSE}\left(\frac{1}{n}+\frac{(x-\overline{x})^2}{S_{xx}}\right) \tag{3.43}$$

Based on Eq. (3.36), when the variances given in Eqs. (3.42) and (3.43) are divided by σ^2 each has a chi-square distribution with $n - 2$ degrees of freedom. We shall use this information to determine the confidence intervals on the estimates of the regression coefficients $\widehat{\beta}_0$ and $\widehat{\beta}_1$ and the fitted line $\widehat{y}$. Prior to doing so, we introduce the definition of the *standard error* for a statistical quantity $\widehat{\theta}$, which is denoted $\text{se}\left(\widehat{\theta}\right)$. The standard error is the positive square root of the variance; thus,

$$\text{se}\left(\widehat{\theta}\right) = \sqrt{\text{Var}\left(\widehat{\theta}\right)} \tag{3.44}$$

where, in the present case, $\widehat{\theta} = \widehat{\beta}_0, \widehat{\beta}_1, \text{or } \widehat{y}$.

We shall now use these results to obtain the respective confidence interval of these three quantities. Recalling the discussion concerning Eq. (2.69), we form the following test statistic for $\widehat{\theta} = \widehat{\beta}_0, \widehat{\beta}_1, \text{or } \widehat{y}$ as

$$T = \frac{\widehat{\theta} - \theta}{\text{se}(\widehat{\theta})} \tag{3.45}$$

Each of these test statistics has $n - 2$ degrees of freedom. Then, using Eq. (2.78) as a guide, the two-sided confidence interval for the statistic $\widehat{\theta}$ is

$$\widehat{\theta} - t_{\alpha/2,n-2}\text{se}(\widehat{\theta}) \le \theta \le \widehat{\theta} + t_{\alpha/2,n-2}\text{se}(\widehat{\theta}) \tag{3.46}$$

which is the $100(1 - \alpha)\%$ *confidence interval* of θ. The explicit expressions for the confidence intervals for β_0, β_1, and $\beta_0 + \beta_1 x$ are given in Table 3.2.

Table 3.2 Summary of simple linear regression relations and the confidence intervals of their fitted parameters[a]

Definitions (data in the form $\{(x_1,y_1), (x_2,y_2), \ldots,(x_n,y_n)\}$)

$$\widehat{y} = \widehat{\beta}_o + \widehat{\beta}_1 x = \overline{y} + \widehat{\beta}_1(x - \overline{x}) \qquad \widehat{\beta}_1 = \frac{S_{xy}}{S_{xx}} \qquad \widehat{\beta}_0 = \overline{y} - \widehat{\beta}_1\overline{x}$$

$$\overline{y} = \frac{1}{n}\sum_{i=1}^{n} y_i \quad \overline{x} = \frac{1}{n}\sum_{i=1}^{n} x_i \quad S_{xx} = \sum_{i=1}^{n}(x_i - \overline{x})^2 \quad S_{xy} = \sum_{i=1}^{n} y_i(x_i - \overline{x})$$

$$\widehat{\sigma}^2 = \frac{SS_E}{n-2} \quad SS_E = \sum_{i=1}^{n}(y_i - \widehat{y}_i)^2 = \sum_{i=1}^{n} e_i^2 \quad SS_R = \sum_{i=1}^{n}(\widehat{y}_i - \overline{y}_i)^2 \quad SS_T = \sum_{i=1}^{n}(y_i - \overline{y})^2$$

$\widehat{\theta}$	$E(\widehat{\theta})$	$\mathrm{Var}(\widehat{\theta})$	$\mathrm{se}(\widehat{\theta})$	Confidence interval
$\widehat{\beta}_0$	β_0	$\widehat{\sigma}^2\left(\frac{1}{n} + \frac{\overline{x}^2}{S_{xx}}\right)$	$\sqrt{\mathrm{Var}(\widehat{\beta}_0)}$	$\widehat{\beta}_0 - t_{\alpha/2,n-2}\mathrm{se}(\widehat{\beta}_0) \le \beta_0 \le \widehat{\beta}_0 + t_{\alpha/2,n-2}\mathrm{se}(\widehat{\beta}_0)$
$\widehat{\beta}_1$	β_1	$\frac{\widehat{\sigma}^2}{S_{xx}}$	$\sqrt{\mathrm{Var}(\widehat{\beta}_1)}$	$\widehat{\beta}_1 - t_{\alpha/2,n-2}\mathrm{se}(\widehat{\beta}_1) \le \beta_1 \le \widehat{\beta}_1 + t_{\alpha/2,n-2}\mathrm{se}(\widehat{\beta}_1)$
$\widehat{y}(x)$	$\beta_0 + \beta_1 x$	$\widehat{\sigma}^2\left(\frac{1}{n} + \frac{(x - \overline{x})^2}{S_{xx}}\right)$	$\sqrt{\mathrm{Var}(\widehat{y}(x))}$	$\widehat{\beta}_0 + \widehat{\beta}_1 x_0 - t_{\alpha/2,n-2}\mathrm{se}(\widehat{y}(x_0)) \le \beta_o + \beta_1 x_0 \le \widehat{\beta}_0 + \widehat{\beta}_1 x_0 + t_{\alpha/2,n-2}\mathrm{se}(\widehat{y}(x_0))$

[a]Many of these relations are numerically evaluated in interactive graphic IG3-2

We are always interested in knowing whether $\widehat{\beta}_1 = 0$. If $\widehat{\beta}_1 = 0$, then y is independent of x. Therefore, the null hypothesis is H_0: $\widehat{\beta}_1 = 0$ and the alternative hypothesis is H_A: $\widehat{\beta}_1 \neq 0$. Then, the test statistic is obtained as

$$T_0 = \frac{\widehat{\beta}_1}{\mathrm{se}(\widehat{\beta}_1)} \tag{3.47}$$

The quantity $\mathrm{se}(\widehat{\beta}_1)$ is given in Table 3.2 and $\widehat{\beta}_1$ is determined from Eq. (3.11). From Case 2 of Table 2.3, the null hypothesis is rejected when either $T_0 < -t_{\alpha/2,n-2}$ or $T_0 > t_{\alpha/2,n-2}$.

If we square T_0 given by Eq. (3.47), then we obtain

Table 3.3 Analysis of variance for simple linear regression[a]

Source of variation	Degrees of freedom	Sum of squares	Mean square	F value	p-value
Regression	1	SS_R	$\text{MSR} = SS_R$	$F_0 = \dfrac{\text{MSR}}{\text{MSE}}$	$\Psi_f(F_0,1,n-2)$
Error	$n-2$	SS_E	$\text{MSE} = \dfrac{SS_E}{n-2}$		
Total	$n-1$	SS_T			

[a]For a set of numerical values, this table is obtained with interactive graphic IG3-2

$$T_0^2 = \frac{\widehat{\beta}_1^2}{\text{Var}\left(\widehat{\beta}_1\right)} = \frac{S_{xx}\widehat{\beta}_1^2}{\text{MSE}} = \frac{\text{MSR}}{\text{MSE}} = F_0 \tag{3.48}$$

where we have used Eqs. (3.30), (3.37), and (3.42). If the numerator and denominator of Eq. (3.48) are divided by σ^2, then the numerator has a chi-square distribution with one degree of freedom and the denominator has a chi-square distribution with $n-2$ degrees of freedom. Referring to Eq. (2.100), we see that the ratio MSR/MSE is an f statistic, which has been denoted F_0. Associated with F_0 is a p-value given by Eq. (2.120), which in this case is

$$p\text{-value} = \Psi_f(F_0, 1, n-2)$$

It is mentioned that Eq. (2.48) where $F_0 = T_0^2$ is only applicable for this case. It does not apply in general.

Several of the previous results are summarized in an *analysis of variance table* as shown in Table 3.3. In this case, the null hypothesis H_0: $\widehat{\beta}_1 = 0$ is typically rejected when the p-value < 0.01.

Example 3.1

We illustrate these results with the data shown in Table 3.4. For these data, we use interactive graphic IG3-2, which uses the relations appearing in Table 3.2, and determine that

$$\begin{aligned} &\bar{y} = 52.18, \quad \bar{x} = 2.792, \quad S_{xx} = 2.724, \quad S_{xy} = 10.18 \\ &\widehat{\beta}_1 = 3.737, \quad \widehat{\beta}_0 = 41.75, \quad \widehat{\sigma}^2 = 0.2951 \\ &SS_T = 43.339, \quad SS_R = 38.027, \quad SS_E = 5.312 \end{aligned}$$

Additionally, we obtain from interactive graphic IG3-2 the results shown in Fig. 3.4 and the analysis of variance given in Table 3.5.

If we are interested in a confidence level of 95%, then we set $\alpha = 0.05$ and find from interactive graphic IG2-2 that $t_{0.025,18} = 2.10$. To test the null hypothesis

Table 3.4 Data used to obtain a simple linear regression fit in Example 3.1

i	x_i	y_i	i	x_i	y_i
1	2.38	51.11	11	2.78	52.87
2	2.44	50.63	12	2.7	52.36
3	2.7	51.82	13	2.36	51.38
4	2.98	52.97	14	2.42	50.87
5	3.32	54.47	15	2.62	51.02
6	3.12	53.33	16	2.8	51.29
7	2.14	49.9	17	2.92	52.73
8	2.86	51.99	18	3.04	52.81
9	3.5	55.81	19	3.26	53.59
10	3.2	52.93	20	2.3	49.77

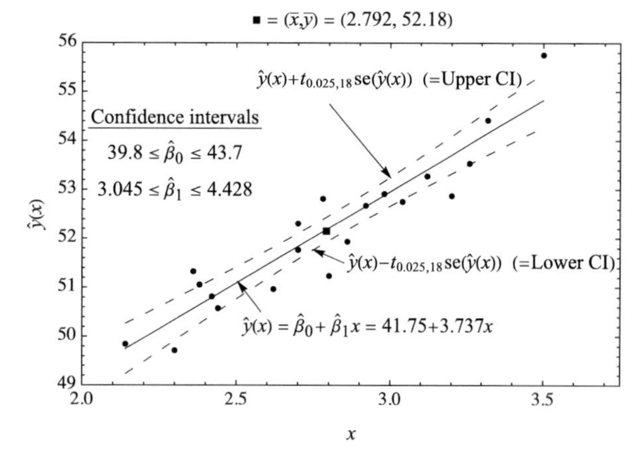

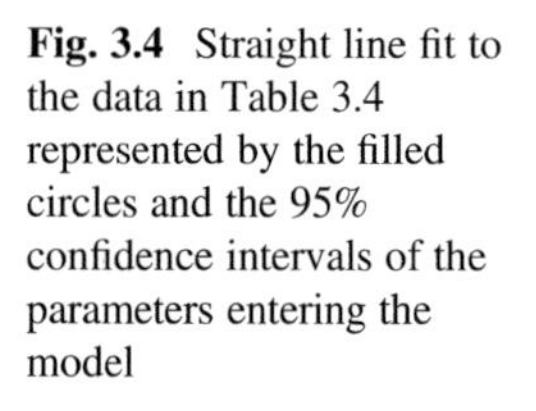

Fig. 3.4 Straight line fit to the data in Table 3.4 represented by the filled circles and the 95% confidence intervals of the parameters entering the model

Table 3.5 Analysis of variance for Example 3.1

Source of variation	Degrees of freedom	Sum of squares	Mean square	F value	p-value
Regression	1	38.03	MSR = 38.03	$F_0 = 128.83$	1.23×10^{-9}
Error	18	5.312	MSE = 0.2951		
Total	19	43.34			

H_0: $\widehat{\beta}_1 = 0$ with the alternative hypothesis H_A: $\widehat{\beta}_1 \neq 0$, we first determine that $\text{se}\left(\widehat{\beta}_1\right) = 0.3292$. Then, from Eq. (3.47), $T_0 = 3.737/0.3292 = 11.35 > 2.10$. Therefore, the null hypothesis is rejected. In terms of the p-value, we find from Eq. (3.47) that $F_0 = T_0^2 = 128.8$ and using Eq. (2.120) and interactive graphic IG2-2 we find that $p = 1.23 \times 10^{-9}$, again indicating that the null hypothesis should be rejected.

From Table 3.2, it is seen that the confidence interval for $y(x)$ varies depending on the value of x. For example, when $x = \overline{x} = 2.792$, which will give the smallest confidence interval, we find that $\widehat{y}(\overline{x}) = 41.75 + 3.737 \times 2.792 = 52.18$ and

$$\mathrm{se}(\widehat{y}(\overline{x})) = \frac{\widehat{\sigma}}{\sqrt{n}} = \frac{0.5432}{4.472} = 0.1215$$

Therefore,

$$\widehat{\beta}_0 + \widehat{\beta}_1\overline{x} - t_{0.025,18}\mathrm{se}(\widehat{y}(\overline{x})) \le \beta_0 + \beta_1\overline{x} \le \widehat{\beta}_0 + \widehat{\beta}_1\overline{x} - t_{0.025,18}\mathrm{se}(\widehat{y}(\overline{x}))$$

which yields

$$\begin{aligned} 52.18 - 2.10 \times 0.1215 \le \beta_0 + \beta_1\overline{x} \le 52.18 + 2.10 \times 0.1215 \\ 51.919 \le y(\overline{x}) \le 52.429 \end{aligned}$$

or $y(\overline{x}) = 52.18 \pm 0.255$. On the other hand, when $x = 3.0$, then

$$\mathrm{se}(\widehat{y}(3.0)) = 0.5432\sqrt{\frac{1}{20} + \frac{(3 - 2.792)^2}{5.321}} = 0.131$$

and we find that $y(3.0) = 52.951 \pm 0.275$.

3.2.3 *Analysis of Residuals*

Analyzing the residuals given by Eq. (3.14) provides a means to confirm that the straight-line fit is appropriate. The residuals e_i can be analyzed three ways. In the first way, they are plotted as a function of the independent variable x_i. From this type of plot, one looks for discernable trends or shapes. If there are none, then most likely the straight-line fit is appropriate.

In the second way, a probability plot of the residuals using the method that resulted in Fig. 2.7 is obtained. This provides a visual means of determining if the residuals are normally distributed. In this case, the residuals are ordered such that the set of residuals $S_e = \{e_1, e_2, \ldots, e_n\}$ becomes $\widehat{S}_e = \{\widetilde{e}_1, \widetilde{e}_2, \ldots, \widetilde{e}_n\}$ where $\widetilde{e}_{i+1} > \widetilde{e}_i$, $i = 1, 2, \ldots, n - 1$. Then,

$$z_i = \frac{\widetilde{e}_i - \overline{e}}{s_e}$$

where

$$\overline{e} = \frac{1}{n}\sum_{i=1}^{n} e_i$$

$$s_e^2 = \frac{1}{n-1}\sum_{i=1}^{n} (e_i - \overline{e})^2$$

and one uses Eq. (2.41) (or one of Eqs. (2.42)–(2.44)) to determine the corresponding ordinate value.

A third way is to compute the standardized residuals d_i, which are defined as

$$d_i = \frac{e_i}{\widehat{\sigma}} = \frac{e_i}{\sqrt{\text{MSE}}} \tag{3.49}$$

where $\widehat{\sigma}$ is given by Eq. (3.35). If the standard residuals d_i are normally distributed, then it is recalled from Eq. (2.29) that approximately 95% of these standardized residuals should fall within the interval (−2, 2). Those values outside these bounds are often considered outliers.

In the following example, we shall analyze a set of residuals using the first two methods.

Example 3.2

To illustrate two ways in which residuals are displayed, we determine the residuals from the data employed to create Fig. 3.4. The residuals e_i are determined from Eq. (3.14). Using interactive graphic IG3-2, we obtain the results shown in Fig. 3.5. In Fig. 3.5a, it is seen that the residuals appear disorderly about the value zero; that is, no trends or patterns can be discerned. In Fig. 3.5b, the residuals have been displayed as a normal probability plot using the procedure that was employed to obtain Fig. 2.7 and p_i is given by Eq. (2.41). It is seen in Fig. 3.5b that the residuals are normally distributed.

Coefficient of Determination

Consider Eq. (3.28) rewritten as

$$R^2 = \frac{SS_R}{SS_T} = 1 - \frac{SS_E}{SS_T} \tag{3.50}$$

The quantity R^2 is called the *coefficient of determination*. It is the proportion of the total variability in the dependent variable y that is accounted for by the regression line. It is seen that $0 \leq R^2 \leq 1$. The larger R^2 is the smaller is the sum of squares errors, indicating that the data points are closer to the regression line. The coefficient of determination is one indicator of judging the adequacy of the regression line. However, it does not indicate the appropriateness of the straight-line model.

For the straight-line fit shown in Fig. 3.4, it is found that $R^2 = 0.8774$.

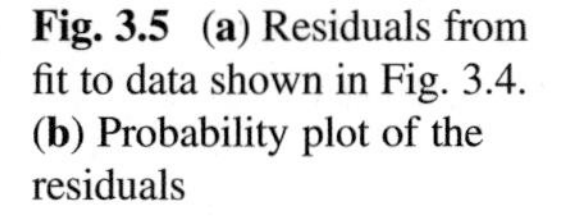

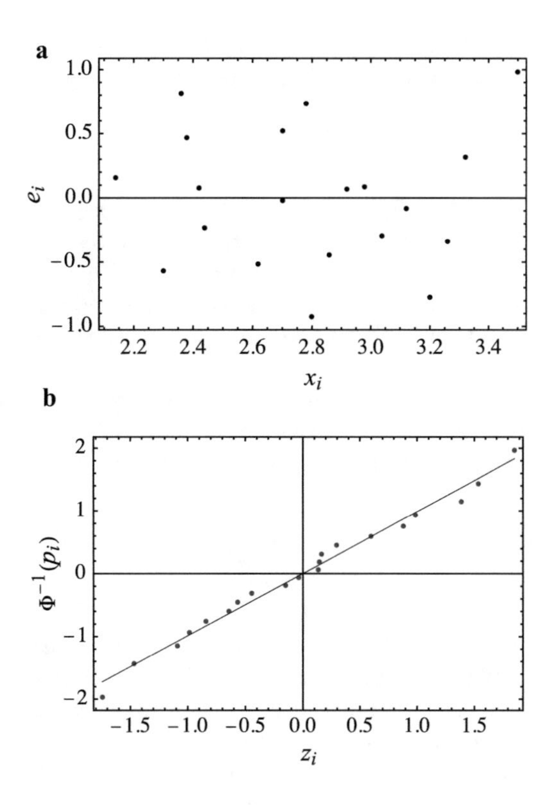

Fig. 3.5 (**a**) Residuals from fit to data shown in Fig. 3.4. (**b**) Probability plot of the residuals

3.3 Multiple Linear Regression

As indicated in the introduction to this chapter, when there is more than one independent variable, the modelling of the data is called *multiple regression*. When the model contains the linear combination of the model's parameter, denoted β_j, $j = 0, 1, \ldots, k$, where $k > 1$, the analysis is called *multiple linear regression* analysis. In this case, the response model is of the form

$$y = \beta_0 + \beta_1 x_1 + \beta_2 x_2 + \cdots + \beta_k x_k = \beta_0 + \sum_{j=1}^{k} \beta_j x_j \tag{3.51}$$

Before proceeding with the determination of β_j, we examine the following models. The first model is a polynomial model

$$y = \beta_0 + \beta_1 x + \beta_2 x^2 + \beta_3 x^4 \tag{3.52}$$

The second model is of the form

$$y = \beta_0 + \beta_1 x_1 + \beta_2 x_2 + \beta_3 x_1^2 + \beta_4 x_2^2 + \beta_5 x_1 x_2 \tag{3.53}$$

Note that the last term in Eq. (3.53) contains the cross product of the independent variables x_1 and x_2. The third model is taken from Eq. (3.1), which is

$$y = \beta_0 + \beta_1 \sin x + \beta_2 x^2 \tag{3.54}$$

Equations (3.52)–(3.54) are multiple linear regression models and each one can be placed in the form given by Eq. (3.51) as follows. If, in Eq. (3.52), we set $x_2 = x^2$ and $x_3 = x^4$, then Eq. (3.52) is in the form of Eq. (3.51). In a similar manner, in Eq. (3.53) we set $x_3 = x_1^2, x_4 = x_2^4$, and $x_5 = x_1 x_2$, and Eq. (3.53) is in the form of Eq. (3.51). Lastly, in Eq. (3.54), we set $x_1 = \sin(x)$ and $x_2 = x^2$ and Eq. (3.54) is of the form of Eq. (3.51).

We continue with Eq. (3.51) with the understanding that, when necessary, the types of variable substitutions indicated for Eqs. (3.52)–(3.54) have taken place.

In a manner analogous to Eq. (3.5), we assume a model of the form

$$y_i = \beta_0 + \sum_{j=1}^{k} \beta_j x_{ij} + \varepsilon_i \quad i = 1, 2, \ldots, n \text{ and } n > k + 1 \tag{3.55}$$

where $\varepsilon_i \sim N(0,\sigma^2)$ is the random error of the ith response. The arithmetic involved in multiple linear regression analysis is onerous and is almost always performed with a computer using any one of many engineering and statistics commercial software programs. Therefore, to minimize the notational complexity of the manipulations, we shall use matrix operations to obtain the results.

The estimates of the β_j, which are denoted $\widehat{\beta}_j, j = 0, 1, \ldots, k$, will be obtained using least squares. Thus, we are interested in obtaining those values of $\widehat{\beta}_j$ that minimize

$$G = \sum_{i=1}^{n} \varepsilon_i^2 = \sum_{i=1}^{n} \left(y_i - \beta_0 - \sum_{j=1}^{k} \beta_j x_{ij} \right)^2 \tag{3.56}$$

Then, the fit model will be

$$\widehat{y} = \widehat{\beta}_0 + \sum_{j=1}^{k} \widehat{\beta}_j x_j \tag{3.57}$$

To place these expressions in matrix notation, we introduced the following vector and matrix quantities

$$\boldsymbol{y} = \begin{Bmatrix} y_1 \\ y_2 \\ \vdots \\ y_n \end{Bmatrix}, \quad \boldsymbol{\varepsilon} = \begin{Bmatrix} \varepsilon_1 \\ \varepsilon_2 \\ \vdots \\ \varepsilon_n \end{Bmatrix}, \quad \boldsymbol{\beta} = \begin{Bmatrix} \beta_0 \\ \beta_1 \\ \vdots \\ \beta_k \end{Bmatrix},$$

$$\boldsymbol{X} = \begin{pmatrix} 1 & x_{11} & x_{12} & \cdots & x_{1k} \\ 1 & x_{21} & x_{22} & \cdots & x_{2k} \\ \vdots & \vdots & \vdots & & \vdots \\ 1 & x_{n1} & x_{n2} & \cdots & x_{nk} \end{pmatrix} \tag{3.58}$$

where $\boldsymbol{y}$ and $\boldsymbol{\varepsilon}$ are $(n \times 1)$ vectors, $\boldsymbol{\beta}$ is an $((k + 1) \times 1)$ vector, and $\boldsymbol{X}$ is an $(n \times (k + 1))$ matrix. The x_{ij} are the inputs with i indicating the run number which produces a y_i and j corresponding to the $k + 1$ independent parameters in the model. The error terms have the vector normal distribution $\boldsymbol{\varepsilon} \sim N(0,\sigma\boldsymbol{I})$, where $\boldsymbol{I}$ is an $(n \times n)$ identity matrix. This normal distribution indicates that

$$E(\boldsymbol{\varepsilon\varepsilon}') = \sigma^2\boldsymbol{I} \tag{3.59}$$

where the prime (′) indicates the transpose of the matrix. Equation (3.59) states that the errors are independent; that is, $\text{cov}(\varepsilon_i, \varepsilon_j) = 0,\ i \neq j$. Then, Eq. (3.55) can be written as

$$\boldsymbol{y} = \boldsymbol{X\beta} + \boldsymbol{\varepsilon} \tag{3.60}$$

and the expected value of $\boldsymbol{y}$ is $E(\boldsymbol{y}) = \boldsymbol{X\beta}$. Placing Eq. (3.56) into matrix notation, we have

$$G = \sum_{i=1}^{n} \varepsilon_i^2 = (\boldsymbol{y} - \boldsymbol{X\beta})'(\boldsymbol{y} - \boldsymbol{X\beta}) \tag{3.61}$$

Expanding Eq. (3.61) and using the properties of the transpose of matrix products,[2] we arrive at

$$\begin{aligned} G &= (\boldsymbol{y}' - \boldsymbol{\beta}'\boldsymbol{X}')(\boldsymbol{y} - \boldsymbol{X\beta}) = \boldsymbol{y}'\boldsymbol{y} - \boldsymbol{\beta}'\boldsymbol{X}'\boldsymbol{y} - \boldsymbol{y}'\boldsymbol{X\beta} + \boldsymbol{\beta}'\boldsymbol{X}'\boldsymbol{X\beta} \\ &= \boldsymbol{y}'\boldsymbol{y} - 2\boldsymbol{y}'\boldsymbol{X\beta} + \boldsymbol{\beta}'\boldsymbol{X}'\boldsymbol{X\beta} \end{aligned} \tag{3.62}$$

To obtain the estimates $\widehat{\beta}_j$, we perform the following matrix operation

[2] Recall that for $(n \times n)$ matrices $\boldsymbol{A}$, $\boldsymbol{B}$, and $\boldsymbol{C}$: $(\boldsymbol{AB})' = \boldsymbol{B}'\boldsymbol{A}'$ and $\boldsymbol{A}'\boldsymbol{B}'\boldsymbol{C} = \boldsymbol{C}'\boldsymbol{BA}$.

$$\frac{\partial G}{\partial \boldsymbol{\beta}} = \frac{\partial}{\partial \boldsymbol{\beta}}(\mathbf{y}'\mathbf{y} - 2\boldsymbol{\beta}'\mathbf{X}'\mathbf{y} + \boldsymbol{\beta}'\mathbf{X}'\mathbf{X}\boldsymbol{\beta}) = -2\mathbf{y}'\mathbf{X} + 2\boldsymbol{\beta}'\mathbf{X}'\mathbf{X} = 0$$

which reduces to

$$\mathbf{X}'\mathbf{X}\widehat{\boldsymbol{\beta}} = \mathbf{X}'\mathbf{y} \tag{3.63}$$

Solving for $\widehat{\boldsymbol{\beta}}$ yields

$$\widehat{\boldsymbol{\beta}} = (\mathbf{X}'\mathbf{X})^{-1}\mathbf{X}'\mathbf{y} \tag{3.64}$$

where the superscript "–1" indicates the matrix inverse. Equation (3.64) represents the k + 1 normal equations for multiple linear regression analysis. Then, the fitted function is

$$\widehat{\mathbf{y}} = \mathbf{X}\widehat{\boldsymbol{\beta}} = \mathbf{H}\mathbf{y} \tag{3.65}$$

where

$$\mathbf{H} = \mathbf{X}(\mathbf{X}'\mathbf{X})^{-1}\mathbf{X}' \tag{3.66}$$

is a symmetric matrix called the 'hat matrix', which transforms the measured values $\mathbf{y}$ into a vector of fitted values $\widehat{\mathbf{y}}$.

The regression residuals are given by

$$\mathbf{e} = \mathbf{y} - \widehat{\mathbf{y}} = \mathbf{y} - \mathbf{X}\widehat{\boldsymbol{\beta}} = \mathbf{y} - \mathbf{H}\mathbf{y} = (\mathbf{I} - \mathbf{H})\mathbf{y} \tag{3.67}$$

The error sum of squares SS_E given in Eq. (3.27) becomes

$$\begin{aligned} SS_E &= \sum_{i=1}^{n} e_i^2 = (\mathbf{y} - \widehat{\mathbf{y}})'(\mathbf{y} - \widehat{\mathbf{y}}) = \left(\mathbf{y} - \mathbf{X}\widehat{\boldsymbol{\beta}}\right)'\left(\mathbf{y} - \mathbf{X}\widehat{\boldsymbol{\beta}}\right) \\ &= \mathbf{y}'\mathbf{y} - \widehat{\boldsymbol{\beta}}'\mathbf{X}'\mathbf{y} - \mathbf{y}'\mathbf{X}\widehat{\boldsymbol{\beta}} + \widehat{\boldsymbol{\beta}}'\mathbf{X}'\mathbf{X}\widehat{\boldsymbol{\beta}} \\ &= \mathbf{y}'\mathbf{y} - 2\widehat{\boldsymbol{\beta}}'\mathbf{X}'\mathbf{y} + \widehat{\boldsymbol{\beta}}'\mathbf{X}'\mathbf{y} \\ &= \mathbf{y}'\mathbf{y} - \widehat{\boldsymbol{\beta}}'\mathbf{X}'\mathbf{y} \end{aligned} \tag{3.68}$$

where we have used Eq. (3.63). Using the form for $\mathbf{e}$ given in Eq. (3.67), we find that SS_E can also be expressed as

$$SS_E = \sum_{i=1}^{n} e_i^2 = \boldsymbol{e}'\boldsymbol{e} = ((\boldsymbol{I} - \boldsymbol{H})\boldsymbol{y})'((\boldsymbol{I} - \boldsymbol{H})\boldsymbol{y})$$
$$= \boldsymbol{y}'(\boldsymbol{I} - \boldsymbol{H})(\boldsymbol{I} - \boldsymbol{H})\boldsymbol{y} = \boldsymbol{y}'(\boldsymbol{I} - \boldsymbol{H})\boldsymbol{y} \tag{3.69}$$

since $\boldsymbol{II} = \boldsymbol{I}$, $\boldsymbol{H}' = \boldsymbol{H}$, and

$$\boldsymbol{HH} = \left(\boldsymbol{X}(\boldsymbol{X}'\boldsymbol{X})^{-1}\boldsymbol{X}'\right)\left(\boldsymbol{X}(\boldsymbol{X}'\boldsymbol{X})^{-1}\boldsymbol{X}'\right) = \boldsymbol{X}(\boldsymbol{X}'\boldsymbol{X})^{-1}\left(\boldsymbol{X}'\boldsymbol{X}(\boldsymbol{X}'\boldsymbol{X})^{-1}\right)\boldsymbol{X}'$$
$$= \boldsymbol{X}(\boldsymbol{X}'\boldsymbol{X})^{-1}\boldsymbol{I}\boldsymbol{X}' = \boldsymbol{X}(\boldsymbol{X}'\boldsymbol{X})^{-1}\boldsymbol{X}' = \boldsymbol{H}$$

Following Eq. (3.35), an unbiased estimate of the variance σ^2 for the multiple regression analysis is

$$\widehat{\sigma}^2 = \frac{SS_E}{n - m} = \frac{SS_E}{n - (k + 1)} \tag{3.70}$$

where m is the number of parameters (the number of β_j) in the model. Recall that for the simple linear regression model, $m = 2$; that is, there were two parameters in the model: β_0 and β_1. In the present notation, $m = k + 1$. From Eq. (2.62), we see that

$$\frac{(n - k - 1)\widehat{\sigma}^2}{\sigma^2} = \frac{SS_E}{\sigma^2} \sim \chi^2_{n-k-1}$$

Then, from Eq. (2.65), the $100(1 - \alpha)\%$ confidence interval is

$$\frac{(n - k - 1)\widehat{\sigma}^2}{\chi^2_{\alpha/2,n-k-1}} \leq \sigma^2 \leq \frac{(n - k - 1)\widehat{\sigma}^2}{\chi^2_{1-\alpha/2,n-k-1}} \tag{3.71}$$

The expected value of $\widehat{\boldsymbol{\beta}}$ is, upon using Eqs. (3.60) and (3.64)

$$E\left(\widehat{\boldsymbol{\beta}}\right) = E\left((\boldsymbol{X}'\boldsymbol{X})^{-1}\boldsymbol{X}'\boldsymbol{y}\right) = E\left((\boldsymbol{X}'\boldsymbol{X})^{-1}\boldsymbol{X}'(\boldsymbol{X}\boldsymbol{\beta} + \boldsymbol{\varepsilon})\right)$$
$$= (\boldsymbol{X}'\boldsymbol{X})^{-1}(\boldsymbol{X}'\boldsymbol{X})\boldsymbol{\beta} + (\boldsymbol{X}'\boldsymbol{X})^{-1}\boldsymbol{X}'E(\boldsymbol{\varepsilon}) \tag{3.72}$$
$$= \boldsymbol{\beta}$$

since

$$(\boldsymbol{X}'\boldsymbol{X})^{-1}(\boldsymbol{X}'\boldsymbol{X}) = \boldsymbol{I} \tag{3.73}$$

and it has been assumed that $E(\boldsymbol{\varepsilon}) = 0$. Therefore, from Eq. (1.36), we see that $\widehat{\boldsymbol{\beta}}$ is an unbiased estimate of $\boldsymbol{\beta}$.

To determine $\mathrm{Var}\left(\widehat{\boldsymbol{\beta}}\right)$, we first note that

$$
\begin{aligned}
\widehat{\boldsymbol{\beta}}-\boldsymbol{\beta} &= (\boldsymbol{X}'\boldsymbol{X})^{-1}\boldsymbol{X}'\boldsymbol{y}-\boldsymbol{\beta}=(\boldsymbol{X}'\boldsymbol{X})^{-1}\boldsymbol{X}'(\boldsymbol{X}\boldsymbol{\beta}+\boldsymbol{\varepsilon})-\boldsymbol{\beta} \\
&= (\boldsymbol{X}'\boldsymbol{X})^{-1}(\boldsymbol{X}'\boldsymbol{X})\boldsymbol{\beta}+(\boldsymbol{X}'\boldsymbol{X})^{-1}\boldsymbol{X}'\boldsymbol{\varepsilon}-\boldsymbol{\beta} \\
&= (\boldsymbol{X}'\boldsymbol{X})^{-1}\boldsymbol{X}'\boldsymbol{\varepsilon}
\end{aligned} \tag{3.74}
$$

Then, using Eqs. (3.59), (3.73), and (3.74), the variance of $\widehat{\boldsymbol{\beta}}$ is

$$
\begin{aligned}
\mathrm{Var}\left(\widehat{\boldsymbol{\beta}}\right) &= E\left(\left(\widehat{\boldsymbol{\beta}}-E\left(\widehat{\boldsymbol{\beta}}\right)\right)\left(\widehat{\boldsymbol{\beta}}-E\left(\widehat{\boldsymbol{\beta}}\right)\right)'\right)=E\left(\left(\widehat{\boldsymbol{\beta}}-\boldsymbol{\beta}\right)\left(\widehat{\boldsymbol{\beta}}-\boldsymbol{\beta}\right)'\right) \\
&= E\left(\left((\boldsymbol{X}'\boldsymbol{X})^{-1}\boldsymbol{X}'\boldsymbol{\varepsilon}\right)\left(\boldsymbol{\varepsilon}'\boldsymbol{X}(\boldsymbol{X}'\boldsymbol{X})^{-1}\right)\right) \\
&= (\boldsymbol{X}'\boldsymbol{X})^{-1}\boldsymbol{X}'E(\boldsymbol{\varepsilon}\boldsymbol{\varepsilon}')\boldsymbol{X}(\boldsymbol{X}'\boldsymbol{X})^{-1} \\
&= (\boldsymbol{X}'\boldsymbol{X})^{-1}\boldsymbol{X}'\sigma^2\boldsymbol{I}\boldsymbol{X}(\boldsymbol{X}'\boldsymbol{X})^{-1}=\sigma^2\boldsymbol{C}
\end{aligned} \tag{3.75}
$$

where

$$
\boldsymbol{C}=(\boldsymbol{X}'\boldsymbol{X})^{-1} \tag{3.76}
$$

The matrix $\boldsymbol{C}$ is a $(k + 1) \times (k + 1)$ matrix. The quantity $\sigma^2\boldsymbol{C}$ is called the *covariance matrix*. Thus, from Eqs. (3.72) and (3.75), we see that $\widehat{\boldsymbol{\beta}} \sim N(\boldsymbol{\beta}, \sigma^2\boldsymbol{C})$. Then, the diagonal terms of $\boldsymbol{C}$ are

$$
\mathrm{Var}\left(\widehat{\beta}_i\right)=\sigma^2C_{ii} \quad i=0,1,2,\ldots,k \tag{3.77}
$$

and the off-diagonal terms are

$$
\mathrm{Cov}\left(\widehat{\beta}_i,\widehat{\beta}_j\right)=\sigma^2C_{ij} \quad i\neq j \tag{3.78}
$$

When σ^2 is not known, which is typically the case, one uses its estimated value given by Eq. (3.70).

We use these results to construct a t statistic as

$$
T=\frac{\widehat{\beta}_i-\beta_i}{\mathrm{se}\left(\widehat{\beta}_i\right)} \tag{3.79}
$$

where, upon using Eq. (3.44),

$$
\mathrm{se}\left(\widehat{\beta}_i\right)=\widehat{\sigma}\sqrt{C_{ii}} \tag{3.80}
$$

and we have replaced σ^2 with its unbiased estimate $\widehat{\sigma}^2$ given by Eq. (3.70). The quantity $\mathrm{se}\left(\widehat{\beta}_i\right)$ is called the *estimated standard error* of $\widehat{\beta}_i$. Then the confidence interval at the $100(1 - \alpha)\%$ confidence level is

$$\widehat{\beta}_i - t_{\alpha/2,n-k-1}\text{se}\left(\widehat{\beta}_i\right) \le \beta_i \le \widehat{\beta}_i + t_{\alpha/2,n-k-1}\text{se}\left(\widehat{\beta}_i\right) \quad i = 0, 1, 2, \ldots, k \tag{3.81}$$

One of the important aspects of multiple regression analysis is the choice and form of the independent variables. In some cases, it may be better to use additional or different independent variables and in other cases it may be better to eliminate one or more of them. One test that helps decide this after the fact is to perform individual hypothesis tests on β_i, $i = 1, 2, \ldots, k$ to determine their importance. In other words, we test the null hypothesis H_0: $\widehat{\beta}_i = 0$ with the alternative hypothesis H_A: $\widehat{\beta}_i \neq 0$. If H_0 is not rejected, the corresponding x_i is most likely not significant with respect to the other regressors in the model and can be removed because it explains a statistically small portion in the variation of $\widehat{y}$. For this hypothesis test, the p-value $= 2\Psi_t(T,n-k-1)$ where T is given by Eq. (3.79) with β_i, $= 0$ and Ψ_t is given by Eq. (2.74).

We shall obtain an estimate of the value of the response for a given set of input values $\boldsymbol{x}_i$, where

$$\boldsymbol{x}_i = \{1 \;\; x_{i1} \;\; x_{i2} \;\; \cdots \;\; x_{ik}\} \tag{3.82}$$

is a $1 \times (k + 1)$ vector that produces the output $\widehat{y}_i$. In Eq. (3.82), the '1' has been placed in the first column of the column vector so that it is compatible with the previous matrices and their subsequent matrix operations. Upon substituting Eq. (3.82) into Eq. (3.65), we obtain

$$\widehat{y}_i = \boldsymbol{x}_i\widehat{\boldsymbol{\beta}}$$

which is a scalar. To determine the variance of $\widehat{y}_i$, we first note from Eq. (3.74) that

$$\boldsymbol{x}_i\left(\widehat{\boldsymbol{\beta}} - \boldsymbol{\beta}\right) = \boldsymbol{x}_i(\boldsymbol{X}'\boldsymbol{X})^{-1}\boldsymbol{X}'\boldsymbol{\varepsilon}$$

Then, using Eqs. (3.75) and (3.59), we find that

$$\begin{aligned}
\text{Var}(\widehat{y}_i) &= \text{Var}\left(\boldsymbol{x}_i'\widehat{\boldsymbol{\beta}}\right) = E\left(\boldsymbol{x}_i\left(\widehat{\boldsymbol{\beta}} - \boldsymbol{\beta}\right)\left(\boldsymbol{x}_i\left(\widehat{\boldsymbol{\beta}} - \boldsymbol{\beta}\right)\right)'\right) \\
&= E\left(\left(\boldsymbol{x}_i(\boldsymbol{X}'\boldsymbol{X})^{-1}\boldsymbol{X}'\boldsymbol{\varepsilon}\right)\left(\boldsymbol{x}_i(\boldsymbol{X}'\boldsymbol{X})^{-1}\boldsymbol{X}'\boldsymbol{\varepsilon}\right)'\right) \\
&= E\left(\boldsymbol{x}_i(\boldsymbol{X}'\boldsymbol{X})^{-1}\boldsymbol{X}'\boldsymbol{\varepsilon}\left(\boldsymbol{\varepsilon}'\boldsymbol{X}(\boldsymbol{X}'\boldsymbol{X})^{-1}\boldsymbol{x}_i'\right)\right) \\
&= \boldsymbol{x}_i(\boldsymbol{X}'\boldsymbol{X})^{-1}\boldsymbol{X}'E(\boldsymbol{\varepsilon}\boldsymbol{\varepsilon}')\boldsymbol{X}(\boldsymbol{X}'\boldsymbol{X})^{-1}\boldsymbol{x}_i' \\
&= \boldsymbol{x}_i\left(\sigma^2(\boldsymbol{X}'\boldsymbol{X})^{-1}\right)\boldsymbol{x}_i' = \boldsymbol{x}_i\text{Var}\left(\widehat{\boldsymbol{\beta}}\right)\boldsymbol{x}_i'
\end{aligned}$$

Consequently,

$$\mathrm{Var}(\widehat{y}_i) = \sigma_i^2 \tag{3.83}$$

where

$$\sigma_i = \widehat{\sigma}\sqrt{\boldsymbol{x}_i(\boldsymbol{X}'\boldsymbol{X})^{-1}\boldsymbol{x}_i'}$$

and we have replaced σ^2 with its unbiased estimate $\widehat{\sigma}^2$ given by Eq. (3.70). The quantity σ_i is sometimes referred to as the *standard error of prediction*. The t statistic can be constructed as

$$T = \frac{\widehat{y}_i - \mu_i}{\sqrt{\mathrm{Var}(\widehat{y}_i)}} = \frac{\widehat{y}_i - \mu_i}{\sigma_i} \tag{3.84}$$

where μ_i is the mean response at $\boldsymbol{x}_i$. The confidence interval at the $100(1-\alpha)\%$ confidence level is

$$\widehat{y}_i - t_{\alpha/2,n-k-1}\sigma_i \le \mu_i \le \widehat{y}_i + t_{\alpha/2,n-k-1}\sigma_i \tag{3.85}$$

These results are summarized in Table 3.6 and in an *analysis of variance table* as shown in Table 3.7. In this case, the null hypothesis is H_0: $\widehat{\beta}_1 = \widehat{\beta}_2 = \cdots = \widehat{\beta}_k = 0$ and the alternative hypothesis is that at least one regressor is important (not zero).

Table 3.6 Summary of multiple linear regression analysis relations and the confidence intervals of the fit parameters

Definitions

$$\boldsymbol{y} = \begin{Bmatrix} y_1 \\ y_2 \\ \vdots \\ y_n \end{Bmatrix}, \quad \boldsymbol{\beta} = \begin{Bmatrix} \beta_0 \\ \beta_1 \\ \vdots \\ \beta_k \end{Bmatrix}, \quad \boldsymbol{x}_i' = \begin{Bmatrix} 1 \\ x_{i1} \\ \vdots \\ x_{ik} \end{Bmatrix}, \quad \boldsymbol{X} = \begin{pmatrix} 1 & x_{11} & x_{12} & \cdots & x_{1k} \\ 1 & x_{21} & x_{22} & \cdots & x_{2k} \\ \vdots & \vdots & \vdots & & \vdots \\ 1 & x_{n1} & x_{n2} & \cdots & x_{nk} \end{pmatrix}$$

$$\widehat{\boldsymbol{y}} = \boldsymbol{X}\widehat{\boldsymbol{\beta}} = \boldsymbol{H}\boldsymbol{y} \qquad \widehat{\boldsymbol{\beta}} = (\boldsymbol{X}'\boldsymbol{X})^{-1}\boldsymbol{X}'\boldsymbol{y} \qquad \boldsymbol{H} = \boldsymbol{X}(\boldsymbol{X}'\boldsymbol{X})^{-1}\boldsymbol{X}'$$

$$\boldsymbol{e} = \boldsymbol{y} - \widehat{\boldsymbol{y}} = \boldsymbol{y} - \boldsymbol{X}\widehat{\boldsymbol{\beta}} = (\boldsymbol{I} - \boldsymbol{H})\boldsymbol{y} \qquad \boldsymbol{C} = (\boldsymbol{X}'\boldsymbol{X})^{-1}$$

$$\mathrm{se}(\widehat{\beta}_i) = \widehat{\sigma}\sqrt{C_{ii}} \quad \sigma_i = \widehat{\sigma}\sqrt{\boldsymbol{x}_i(\boldsymbol{X}'\boldsymbol{X})^{-1}\boldsymbol{x}_i'} \quad \widehat{\sigma}^2 = \frac{\boldsymbol{y}'\boldsymbol{y} - \widehat{\boldsymbol{\beta}}'\boldsymbol{X}'\boldsymbol{y}}{n-(k+1)} = \frac{\boldsymbol{y}'(\boldsymbol{I}-\boldsymbol{H})\boldsymbol{y}}{n-(k+1)}$$

$\widehat{\theta}$	$E(\widehat{\theta})$	$\mathrm{Var}(\widehat{\theta})$	Confidence interval ($i = 0,1, 2, \ldots, k$)
$\widehat{\boldsymbol{\beta}}$	$\boldsymbol{\beta}$	$\sigma^2\boldsymbol{C}$	$\widehat{\beta}_i - t_{\alpha/2,n-k-1}\mathrm{se}(\widehat{\beta}_i) \le \beta_i \le \widehat{\beta}_i + t_{\alpha/2,n-k-1}\mathrm{se}(\widehat{\beta}_i)$
$\widehat{y}_i = \boldsymbol{x}_i\widehat{\boldsymbol{\beta}}$	$y_i = \boldsymbol{x}_i\boldsymbol{\beta}$	σ_i^2	$\widehat{y}_i - t_{\alpha/2,n-k-1}\sigma_i \le \mu_i \le \widehat{y}_i + t_{\alpha/2,n-k-1}\sigma_i$

Table 3.7 Analysis of variance for multiple linear regression analysis, where $a_0 = n - (k + 1)$

Source of variation	Degrees of freedom	Sum of squares[a]	Mean square	F value	p-value
Regression	k	$SS_R = \sum_{i=1}^{n} (\widehat{y}_i - \overline{y})^2 = \mathbf{y}'\left(\boldsymbol{H} - \frac{\boldsymbol{J}}{n}\right)\mathbf{y}$	$\text{MSR} = \frac{SS_R}{k}$	$F_0 = \frac{\text{MSR}}{\text{MSE}}$	$\Psi_f(F_0, k, a_0)$
Error	a_0	$SS_E = \sum_{i=1}^{n} (y_i - \widehat{y}_i)^2 = \mathbf{y}'(\boldsymbol{I} - \boldsymbol{H})\mathbf{y}$	$\text{MSE} = \frac{SS_E}{a_0}$		
Total	$n - 1$	$SS_T = \sum_{i=1}^{n} (y_i - \overline{y})^2 = \mathbf{y}'\left(\boldsymbol{I} - \frac{\boldsymbol{J}}{n}\right)\mathbf{y}$			

[a]The quantity $\boldsymbol{J}$ is an $(n \times n)$ matrix of ones

Table 3.8 Data for Example 3.3

i	Response y_i	Input x_{i1}	Input x_{i2}	Input x_{i3}
1	25.62	1.679	5.496	10.62
2	31.06	6.496	5.614	9.616
3	26.27	6.035	8.371	7.352
4	38.85	10.42	4.437	8.414
5	18.42	1.028	11.54	9.204
6	27.17	1.376	6.03	9.665
7	26.17	3.952	6.784	7.986
8	26.08	6.228	8.884	8.885
9	32.06	4.228	4.288	8.567
10	25.12	4.109	7.583	9.167
11	40.18	10.18	4.763	9.666
12	35.48	1.608	3.287	7.849
13	26.55	1.602	5.404	8.35

Example 3.3

We shall now illustrate the preceding results with the data shown in Table 3.8. For these data, we assume the following model, which contains three independent input variables

$$y = \beta_0 + \beta_1 x_1 + \beta_2 x_2 + \beta_3 x_3$$

The subsequent results were obtained with interactive graphic IG3-3. The solution for the estimated regression coefficients, respective standard error, and p-values are given in Table 3.9. It is seen from Table 3.9 that the p-value for $\widehat{\beta}_3$ is large indicating that it may be excluded from the model. The analysis of variancc is given in Table 3.10. For these data, we find from Eq. (3.50) that $R^2 = 0.9146$ and from Eq. (3.70) that $\widehat{\sigma} = 2.19$.

The fit values, residuals, and confidence intervals at the 95% confidence level for the fit values are displayed in Fig. 3.6, where it is seen that all but one of the observed

Table 3.9 Estimated regression coefficients and their properties for simple linear regression of the data in Table 3.8

β	$\widehat{\beta}$	$se\left(\widehat{\beta}\right)$	T_0	p-value
β_0	38.78	6.324	6.132	0.000172
β_1	1.017	0.1901	5.35	0.000462
β_2	−1.869	0.2693	−6.939	0.0000677
β_3	−0.2682	0.6624	−0.4049	0.695
	Eq. (3.64)	Eq. (3.80)	Eq. (3.79)	Source

Table 3.10 Analysis of variance for simple linear regression of the data in Table 3.8

Source of variation	Degrees of Freedom	Sum of Squares	Mean square	F_0	p-value
Regression	3	410.72	136.91	32.12	0.0000777
Error	9	38.363	4.263		
Total	12	449.08			

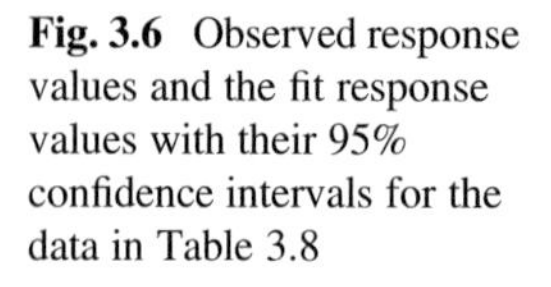

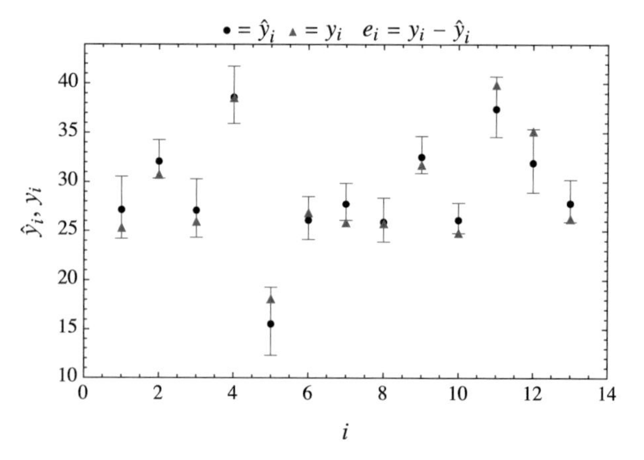

Fig. 3.6 Observed response values and the fit response values with their 95% confidence intervals for the data in Table 3.8

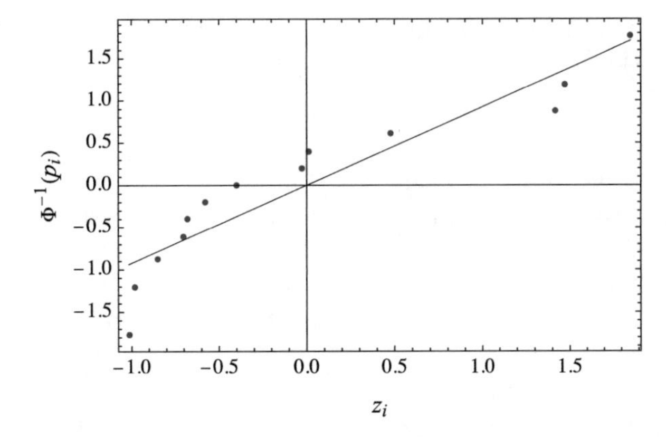

Fig. 3.7 Probability plot of the residuals of the multiple linear regression fit to the data in Table 3.8

response values y_i are within the 95% confidence intervals. A probability plot of the residuals is given in Fig. 3.7, where it is seen that the data are somewhat normally distributed. In obtaining Fig. 3.7, Eq. (2.41) was used.

3.4 Exercises

Section 3.2

3.1. For the data in Table 3.11, determine which of the transformations given by cases 1, 2, or 3 in Table 3.1 provide the best fit to these data. Justify your answer.

3.2. For the data shown in Table 3.12, use a straight-line model and:

(a) Determine the coefficients of the model and their confidence interval at the 95% level
(b) Determine the coefficient of determination
(c) Construct a scatter plot and probability plot of the residuals
(d) Summarize the results in an ANOVA table and determine that it is reasonable to assume that $\beta_1 \neq 0$.
(e) Determine if there are any outliers based on a qualitative analysis of using Eq. (3.49).
(f) From the results obtained in (a) through (e), justify that the model is an appropriate one.

Table 3.11 Data for Exercise 3.1

x	y
0.25	1.87
0.525	1.48
0.80	1.30
1.08	1.01
1.35	0.86
1.63	0.772
1.90	0.805
2.18	0.734
2.45	0.588
2.73	0.522
3.00	0.572

Table 3.12 Data for Exercise 3.2

x	y
6.715	96.84
7.76	99.05
10.42	96.07
10.8	101.6
11.47	90.41
13.27	92.14
14.98	90.31
15.55	93.1
15.84	89.45
16.69	85.51
18.21	87.14
19.07	86.09
20.02	89.16
20.87	84.55
25.05	83.4

Table 3.13 Data for Exercise 3.3

x	y
2.0	25.985
2.7	28.830
3.4	36.678
4.1	40.445
4.8	45.521
5.5	45.955
6.2	49.800
6.9	51.343
7.6	52.780
8.3	53.897
9.0	54.191

Section 3.3

3.3. For the data shown in Table 3.13 and for $\alpha = 0.05$:

(a) Fit a straight line to these data and from a plot of the fitted line, an ANOVA table, a scatter plot of the residuals, and a probability plot of the residuals what can you conclude.
(b) Fit a quadratic polynomial to these data and from a plot of the fitted curve, an ANOVA table, a scatter plot of the residuals, and a probability plot of the residuals what can you conclude.
(c) Repeat part (b) using a cubic polynomial fit to the data.
(d) Which is the better fit to these data: straight line, quadratic, or cubic? Justify your choice.

Chapter 4
Experimental Design

In this chapter, the terms used in experimental design are introduced: response variable, factor, extraneous variable, level, treatment, blocking variable, replication, contrasts, and effects. The relations needed to analyze a one-factor experiment, a randomized complete block design, a two-factor experiment, and a 2^k-factorial experiment are derived. For these experiments, an analysis of variance is used to determine the factors that are most influential in determining its output and, where appropriate, whether the factors interact.

4.1 Introduction

In engineering, experiments are performed to discover something about a particular process with the goal of improving it or understanding it. The improvement or understanding can be in the process itself or what the process produces. A *designed experiment* is a test or a series of tests in which specific changes are made to the inputs of the process to identify reasons for changes in its output. In a sense, all experiments are designed experiments. It just that some experiments are better designed than others.

Experiments involve the following activities. It starts with a conjecture, which is the original hypothesis that motivates the experiment. This is followed by the experiment that investigates the conjecture. After completion of the experiment, the data are analyzed and conclusions are drawn, which may lead to a revised conjecture. Designed experiments are often used do such things as improve process yield, reduce costs, meet performance metrics, and compare alternative processes and materials.

Supplementary Information The online version contains supplementary material available at [https://doi.org/10.1007/978-3-031-05010-7_4].

E. B. Magrab, *Engineering Statistics*, https://doi.org/10.1007/978-3-031-05010-7_4

The language used in a designed experiment is as follows.

The *response variable* is the measured output of the experiment.

A *factor* is a variable that is varied in a controlled manner to observe its impact on the response variable.

An *extraneous variable* is one that can potentially affect the response variable but is not of interest as a factor. It is also called noise, a nuisance variable, or a blocking variable.

A *level* is a given value of a factor and is often referred to as a *treatment*.

Replication is the number of times that an experiment is repeated at each level.

Randomization means that the levels of a factor and the order in which the individual experimental runs are made are randomly determined. Randomization greatly diminishes the influence of extraneous variables, minimizes bias from the operator and the process, provides a more realistic estimate of the variance, and assures the validity of the probability statements of the statistical tests used to analyze the data. The random assignment of treatments does not a mean haphazard assignment. It means one uses a table of random numbers or a computer program to generate random numbers and then uses these numbers to assign the order in which the treatments are tested.

A few examples of a single factor experiment are:

1. Three production lines making the same product; response variable = specific characteristic of product; factor = production line; levels = specific production line
2. Tensile strength of a material as a function of additives: response variable = tensile strength; factor = additive; levels = specific additive
3. Asphalt resilience for different aggregate: response variable = resilience; factor = aggregate; levels = specific aggregate
4. Carpet wear for different fiber composition: response variable = wear; factor = fiber composition; levels = specific composition

4.2 One-Factor Analysis of Variance

A *one-factor analysis of variance* is used to determine the effect of only one independent variable by comparing the means from three or more populations. When there are two populations, we use the methods discussed for Cases 3 to 5 in Table 2.2.

Consider a factor A that has $A_1, A_2, \ldots, A_a$ levels (treatments). At each level A_i, we record n_i response measurements (observations, replicates). The response measurements are denoted y_{ij}, $i = 1, 2, \ldots, a$ and $j = 1, 2, \ldots, n_i$. The data are organized as shown in Table 4.1. Since the experiment must be performed as a completely randomized design, each observation in Table 4.1 is obtained in a randomly determined order and not in the order that one may infer from its position in the table. That is, for the first run (replicate), the levels are selected randomly. For the second run, a

Table 4.1 Single factor analysis of variance data and their means, variances, and residuals

	Observation (replicate) (j)			Treatment average	Treatment variance	Residual
Treatment Level (i)	1 ⋯	k ⋯	n_i	$\bar{y}_i = \frac{1}{n_i}\sum_{j=1}^{n_i} y_{ij}$	$s_i^2 = \frac{1}{n_i - 1}\sum_{j=1}^{n_i} \Delta_{ij}^2$ $\Delta_{ij} = y_{ij} - \bar{y}_i$	$e_{ij} = y_{ij} - \bar{y}_i$
A_1	y_{11}	y_{1k}	y_{1n_1}	$\bar{y}_1$	s_1^2	$e_{1j} = y_{1j} - \bar{y}_1$
⋮	⋮	⋮	⋮	⋮	⋮	
A_m	y_{m1}	y_{mk}	y_{mn_m}	$\bar{y}_m$	s_m^2	$e_{mj} = y_{mj} - \bar{y}_m$
⋮	⋮	⋮	⋮	⋮	⋮	
A_a	y_{a1}	y_{ak}	y_{an_a}	$\bar{y}_a$	s_a^2	$e_{aj} = y_{aj} - \bar{y}_a$

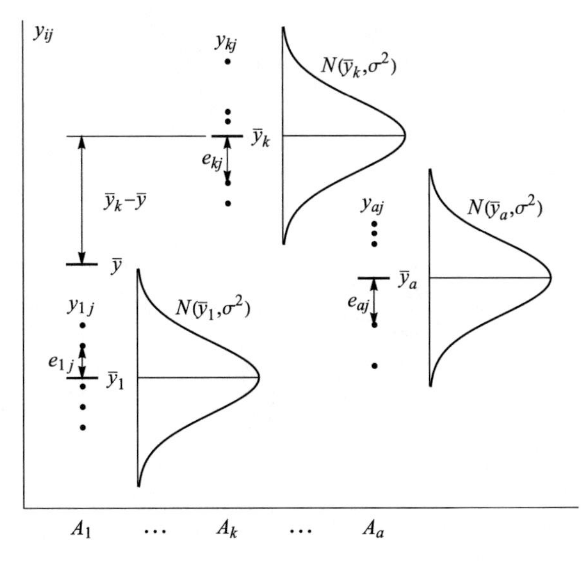

Fig. 4.1 Graphical representation of the data and computed quantities in Table 4.1 for $n_i = 5$, $i = ,1, 2, \ldots, a$, and the corresponding assumed normal distribution

new set of randomly selected numbers are obtained to determine the order of the levels, and so on. The results in Table 4.1 are also shown graphically in Fig. 4.1, where we have for clarity assumed that $n_i = 5$, $i = 1, 2, \ldots, a$. The total number of measurements is

$$n_T = \sum_{i=1}^{a} n_i \tag{4.1}$$

The ith treatment mean and variance, respectively, are given by

$$\begin{aligned}\bar{y}_i &= \frac{1}{n_i}\sum_{j=1}^{n_i} y_{ij} \\ s_i^2 &= \frac{1}{n_i - 1}\sum_{j=1}^{n_i}\left(y_{ij} - \bar{y}_i\right)^2\end{aligned} \tag{4.2}$$

When the n_i are not all equal, the one factor analysis of variance is called *unbalanced* and when they are all equal the analysis is called *balanced*. The balanced analysis of variance is relatively insensitive to small departures from the assumption of equality of variances as discussed next.

It is noted that since the levels A_j were chosen by the user, the conclusions reached from the analysis of the data cannot be extended to other levels. This model is called the *fixed-effects* model.

We now obtain the *sum of squares identity* to partition the variability. Following the procedure indicated by Eqs. (3.25) and (3.26), we have that

$$\begin{aligned}\sum_{i=1}^{a}\sum_{j=1}^{n_i}\left(y_{ij} - \bar{y}\right)^2 &= \sum_{i=1}^{a}\sum_{j=1}^{n_i}\left((\bar{y}_i - \bar{y}) + \left(y_{ij} - \bar{y}_i\right)\right)^2 \\ &= \sum_{i=1}^{a}\sum_{j=1}^{n_i}(\bar{y}_i - \bar{y})^2 + \sum_{i=1}^{a}\sum_{j=1}^{n_i}\left(y_{ij} - \bar{y}_i\right)^2 + c_p\end{aligned}$$

where

$$\begin{aligned}c_p &= 2\sum_{i=1}^{a}\sum_{j=1}^{n_i}(\bar{y}_i - \bar{y})\left(y_{ij} - \bar{y}_i\right) = 2\sum_{i=1}^{a}\left\{(\bar{y}_i - \bar{y})\sum_{j=1}^{n_i}\left(y_{ij} - \bar{y}_i\right)\right\} \\ &= 2\sum_{i=1}^{a}\left\{(\bar{y}_i - \bar{y})\left(\sum_{j=1}^{n_i} y_{ij} - n_i\bar{y}_i\right)\right\} = 2\sum_{i=1}^{a}(\bar{y}_i - \bar{y})(n\bar{y}_i - n\bar{y}_i) \\ &= 0\end{aligned}$$

and

$$\bar{y} = \frac{1}{n_T}\sum_{i=1}^{a}\sum_{j=1}^{n_j} y_{ij} = \frac{1}{n_T}\sum_{i=1}^{a} n_i\bar{y}_i \tag{4.3}$$

is a point estimate of μ. Therefore, the *sum of squares identity* becomes

$$\sum_{i=1}^{a}\sum_{j=1}^{n_i}\left(y_{ij} - \bar{y}\right)^2 = \sum_{i=1}^{a}\sum_{j=1}^{n_i}(\bar{y}_i - \bar{y})^2 + \sum_{i=1}^{a}\sum_{j=1}^{n_i}\left(y_{ij} - \bar{y}_i\right)^2$$

However,

$$\sum_{i=1}^{a}\sum_{j=1}^{n_i}(\bar{y}_i-\bar{y})^2=\sum_{i=1}^{a}\sum_{j=1}^{n_i}(\bar{y}_i^2-2\bar{y}_i\bar{y}+\bar{y}^2)=\sum_{i=1}^{a}n_i\bar{y}_i^2-2\bar{y}\sum_{i=1}^{a}n_i\bar{y}_i+\bar{y}^2\sum_{i=1}^{a}n_i$$
$$=\sum_{i=1}^{a}n_i\left(\bar{y}_i^2-2\bar{y}\bar{y}_i+\bar{y}^2\right)$$
$$=\sum_{i=1}^{a}n_i(\bar{y}_i-\bar{y})^2$$

Then, the final form of the sum of squares identity is written as

$$\sum_{i=1}^{a}\sum_{j=1}^{n_i}\left(y_{ij}-\bar{y}\right)^2=\sum_{i=1}^{a}n_i(\bar{y}_i-\bar{y})^2+\sum_{i=1}^{a}\sum_{j=1}^{n_i}\left(y_{ij}-\bar{y}_i\right)^2 \tag{4.4}$$

Introducing the notation

$$SS_T=\sum_{i=1}^{a}\sum_{j=1}^{n_i}\left(y_{ij}-\bar{y}\right)^2=\sum_{i=1}^{a}\sum_{j=1}^{n_i}y_{ij}^2-n_T\bar{y}^2 \quad \text{(total sum of squares)}$$
$$SS_A=\sum_{i=1}^{a}n_i(\bar{y}_i-\bar{y})^2=\sum_{i=1}^{a}n_i\bar{y}_i^2-n_T\bar{y}^2 \quad \text{(treatment sum of squares)}$$
$$SS_E=\sum_{i=1}^{a}\sum_{j=1}^{n_i}\left(y_{ij}-\bar{y}_i\right)^2=\sum_{i=1}^{a}\sum_{j=1}^{n_i}y_{ij}^2-\sum_{i=1}^{a}n_i\bar{y}_i^2 \quad \text{(error sum of squares)}$$
$$\tag{4.5}$$

Eq. (4.4) can be written as

$$SS_T=SS_A+SS_E \tag{4.6}$$

Thus, the sum of squares SS_T uses the deviations of all measurements from the grand mean; that is, it is a measure of the total variability in the data. The sum of squares SS_A uses the deviations of each level's mean from the grand mean and is a measure of the variability *between* treatment levels. The sum of squares SS_E uses the deviations of all observations within each level from that level's mean and then sums these deviations for all levels. It is a measure of the variability *within* each treatment level. From the definition of an unbiased variance, we see that SS_A has $a-1$ degrees of freedom, SS_T has n_T-1 degrees of freedom, and therefore, SS_E has n_T-a degrees of freedom.

The mean square quantities for the treatments and the errors, respectively, are

$$\begin{aligned}\text{MSA} &= \frac{SS_A}{a-1} \\ \text{MSE} &= \frac{SS_E}{n_T - a}\end{aligned} \tag{4.7}$$

We now determine the expected value of SS_E. To do this, we assume a linear model for the response y_{ij} as

$$y_{ij} = \mu_i + \varepsilon_{ij} = \mu + \tau_i + \varepsilon_{ij} \tag{4.8}$$

where ε_{ij} represents the independent normal random error with $\varepsilon_{ij} \sim N(0,\sigma^2)$, μ_i is the unknown ith treatment mean, μ is the unknown overall (grand) mean, where

$$\mu = \frac{1}{n_T}\sum_{i=1}^{a} n_i\mu_i \tag{4.9}$$

and τ_i is the deviation of the ith treatment mean from the grand mean given by

$$\tau_i = \mu_i - \mu \tag{4.10}$$

It is seen from Eqs. (4.9) and (4.10) that

$$\sum_{i=1}^{a} n_i\tau_i = \sum_{i=1}^{a} n_i(\mu_i - \mu) = \sum_{i=1}^{a} n_i\mu_i - \mu\sum_{i=1}^{a} n_i = n_T\mu - n_T\mu = 0 \tag{4.11}$$

Next, we use Eq. (4.8) in Eq. (4.2) and find that

$$\overline{y}_i = \frac{1}{n_i}\sum_{j=1}^{n_i} y_{ij} = \frac{1}{n_i}\sum_{j=1}^{n_i}\left(\mu + \tau_i + \varepsilon_{ij}\right) = \mu + \tau_i + \overline{\varepsilon}_i \tag{4.12}$$

where

$$\overline{\varepsilon}_i = \frac{1}{n_i}\sum_{j=1}^{n_i} \varepsilon_{ij} \tag{4.13}$$

Then

$$y_{ij} - \overline{y}_i = \mu + \tau_i + \varepsilon_{ij} - \left(\mu + \tau_i + \overline{\varepsilon}_i\right) = \varepsilon_{ij} - \overline{\varepsilon}_i$$

and the expected value of SS_E can be written as

$$\begin{aligned}
E(SS_E) &= E\left(\sum_{i=1}^{a}\sum_{j=1}^{n_i}(y_{ij}-\bar{y}_i)^2\right) = E\left(\sum_{i=1}^{a}\sum_{j=1}^{n_i}(\varepsilon_{ij}-\bar{\varepsilon}_i)^2\right) \\
&= E\left(\sum_{i=1}^{a}\sum_{j=1}^{n_i}\left(\varepsilon_{ij}^2 - 2\varepsilon_{ij}\bar{\varepsilon}_i + \bar{\varepsilon}_i^2\right)\right) \\
&= E\left(\sum_{i=1}^{a}\sum_{j=1}^{n_i}\varepsilon_{ij}^2 - 2\sum_{i=1}^{a}\bar{\varepsilon}_i\sum_{j=1}^{n_i}\varepsilon_{ij} + \sum_{i=1}^{a} n_i\bar{\varepsilon}_i^2\right) \\
&= \sum_{i=1}^{a}\sum_{j=1}^{n} E\left(\varepsilon_{ij}^2\right) - \sum_{i=1}^{a} n_i E\left(\bar{\varepsilon}_i^2\right)
\end{aligned} \tag{4.14}$$

where we have used Eq. (4.13). It was assumed that $\varepsilon_{ij} \sim N(0,\sigma^2)$; therefore, $\bar{\varepsilon}_i \sim N(0, \sigma^2/n_i)$ and

$$E(\varepsilon_{ij}^2) = \sigma^2, \quad E\left(\bar{\varepsilon}_i^2\right) = \frac{\sigma^2}{n_i} \tag{4.15}$$

since their respective mean values are zero. Substituting Eq. (4.15) into Eq. (4.14), we obtain

$$\begin{aligned}
E(SS_E) &= \sum_{i=1}^{a}\sum_{j=1}^{n_i}\sigma^2 - \sum_{i=1}^{a} n_i\left(\frac{\sigma^2}{n_i}\right) = n_T\sigma^2 - a\sigma^2 \\
&= (n_T - a)\sigma^2
\end{aligned}$$

Thus, using Eq. (4.7)

$$E\left(\frac{SS_E}{n_T - a}\right) = E(\text{MSE}) = \sigma^2$$

and an unbiased estimator of the error variance is

$$\widehat{\sigma}^2 = \frac{SS_E}{n_T - a} = \text{MSE} \tag{4.16}$$

which has $(n_T - a)$ degrees of freedom.

To determine the expected value of SS_A, we first note from Eqs. (4.3), (4.8), and (4.11) that

$$\begin{aligned}\bar{y} &= \frac{1}{n_T}\sum_{i=1}^{a}\sum_{j=1}^{n_i} y_{ij} = \frac{1}{n_T}\sum_{i=1}^{a}\sum_{j=1}^{n_i}\left(\mu+\tau_i+\varepsilon_{ij}\right) \\ &= \mu + \frac{1}{n_T}\sum_{i=1}^{a}\sum_{j=1}^{n_i}\tau_i + \frac{1}{n_T}\sum_{i=1}^{a}\sum_{j=1}^{n_i}\varepsilon_{ij} \\ &= \mu + \frac{1}{n_T}\sum_{i=1}^{a} n_i\tau_i + \bar{\varepsilon} = \mu + \bar{\varepsilon}\end{aligned} \tag{4.17}$$

where

$$\bar{\varepsilon} = \frac{1}{n_T}\sum_{i=1}^{a}\sum_{j=1}^{n_i}\varepsilon_{ij} = \frac{1}{n_T}\sum_{i=1}^{a} n_i\bar{\varepsilon}_i \tag{4.18}$$

Then, using Eqs. (4.12) and (4.17),

$$\begin{aligned}(\bar{y}_i-\bar{y})^2 &= (\mu+\tau_i+\bar{\varepsilon}_i-\mu-\bar{\varepsilon})^2 = (\tau_i+\bar{\varepsilon}_i-\bar{\varepsilon})^2 \\ &= \tau_i^2 + 2\tau_i\bar{\varepsilon}_i - 2\bar{\varepsilon}\tau_i + \bar{\varepsilon}_i^2 - 2\bar{\varepsilon}_i\bar{\varepsilon} + \bar{\varepsilon}^2\end{aligned} \tag{4.19}$$

Using Eqs. (4.11), (4.18), and (4.19) in Eq. (4.5), the expected value of SS_A can be written as

$$\begin{aligned}E(SS_A) &= E\left[\sum_{i=1}^{a} n_i(\bar{y}_i-\bar{y})^2\right] = E\left[\sum_{i=1}^{a} n_i\left(\tau_i^2 + 2\tau_i\bar{\varepsilon}_i - 2\bar{\varepsilon}\tau_i + \bar{\varepsilon}_i^2 - 2\bar{\varepsilon}_i\bar{\varepsilon} + \bar{\varepsilon}^2\right)\right] \\ &= E\left[\sum_{i=1}^{a} n_i\tau_i^2 + 2\sum_{i=1}^{a} n_i\tau_i\bar{\varepsilon}_i - 2\bar{\varepsilon}\sum_{i=1}^{a} n_i\tau_i + \sum_{i=1}^{a} n_i\bar{\varepsilon}_i^2 - 2\bar{\varepsilon}\sum_{i=1}^{a} n_i\bar{\varepsilon}_i + \bar{\varepsilon}^2\sum_{i=1}^{a} n_i\right] \\ &= \sum_{i=1}^{a} n_i\tau_i^2 + 2\sum_{i=1}^{a} n_i\tau_i E(\bar{\varepsilon}_i) + \sum_{i=1}^{a} n_i E\left(\bar{\varepsilon}_i^2\right) - n_T E\left(\bar{\varepsilon}^2\right)\end{aligned} \tag{4.20}$$

However, from Eqs. (4.13) and (4.15)

$$E(\bar{\varepsilon}_i) = \frac{1}{n_i}\sum_{j=1}^{n_i} E\left(\varepsilon_{ij}\right) = 0 \qquad E\left(\bar{\varepsilon}_i^2\right) = \frac{\sigma^2}{n_i} \qquad E\left(\bar{\varepsilon}^2\right) = \frac{\sigma^2}{n_T} \tag{4.21}$$

Substituting Eq. (4.21) into Eq. (4.20), we arrive at

$$E(SS_A) = \sum_{i=1}^{a} n_i\tau_i^2 + \sum_{i=1}^{a} n_i\left(\frac{\sigma^2}{n_i}\right) - n_T\left(\frac{\sigma^2}{n_T}\right)$$
$$= (a-1)\sigma^2 + \sum_{i=1}^{a} n_i\tau_i^2$$

or

$$E\left(\frac{SS_A}{a-1}\right) = \sigma^2 + \frac{1}{a-1}\sum_{i=1}^{a} n_i\tau_i^2 \tag{4.22}$$

Thus, the expected value of $SS_A/(a-1)$ is a *biased* estimate of the variance expect when $\tau_i = 0$, $i = 1, 2, \ldots, a$; that is, when all the means $\bar{y}_i$ are equal.[1] The bias is positive so that Eq. (4.22) overestimates the variance. Thus, the unbiased estimator of the error variance given by Eq. (4.16) is considered a good estimate of σ^2 as it is unaffected by τ_i.

As was done with regression analysis, one checks the validity of the assumption that the data are normally distributed by using a probability plot of the residuals, which are $e_{ij} = y_{ij} - \bar{y}_i$.

It is noted that when a balanced analysis is employed, then $n_i = n$ and the previous results simplify in an obvious manner and $n_T = an$.

The above results are used to test the null hypothesis

$$H_0 : \bar{y}_1 = \cdots = \bar{y}_a \tag{4.23}$$

or, equivalently,

$$H_0 : \tau_1 = \cdots = \tau_a = 0 \tag{4.24}$$

with the alternative H_A: $\bar{y}_i$ are not all equal. The test statistic is the F-statistic given by

$$F_0 = \frac{\text{MSA}}{\text{MSE}}$$

with $a - 1$ degrees of freedom in the numerator and $n_T - a$ degrees of freedom in the denominator. The null hypothesis H_0 does not indicate that the treatments do not have any effect, it indicates only that they have the same effect. Therefore, the f statistic provides a test of whether there are differences in the treatment effects and not whether any treatment effects exist.

These results are summarized in Table 4.2.

[1] When $\mu_i = \mu_j$, $i \neq j$, we see from Eq. (4.10) that $\mu + \tau_i = \mu + \tau_j \rightarrow \tau_i = \tau_j$. Then, Eq. (4.11) becomes $\sum_{i=1}^{a} n_i\tau_i = n_T\tau_i = 0$; therefore, τ_i must equal zero for all i.

Table 4.2 Single factor analysis of variance

Source of variation	Degrees of freedom	Sum of squares	Mean square	F value	p-value[a]
Treatment	$a-1$	SS_A	$\text{MSA} = \dfrac{SS_A}{a-1}$	$F_0 = \dfrac{\text{MSA}}{\text{MSE}}$	$\Psi_f(F_0, a-1, n_T-a)$
Error	n_T-a	SS_E	$\text{MSE} = \dfrac{SS_E}{n_T-a}$		
Total	n_T-1	SS_T			

[a] Ψ_f is determined from Eq. (2.120)

As was done with the hypothesis testing for one population to determine whether $\mu = \mu_0$ or for two populations to determine whether $\mu_1 - \mu_2 = \delta_0$, the confidence intervals for μ_i and $\mu_i - \mu_k$ can also be determined in a similar manner. The difference is in the choice for the standard deviation, which for the analysis of variance is that given by Eq. (4.16); that is, $\widehat{\sigma}$. Then, the 100(1 − α)% confidence interval for μ_i is obtained from Eq. (2.78), which, after notational changes, becomes

$$\bar{y}_i - t_{\alpha/2,n_T-a}\widehat{\sigma}/\sqrt{n_i} \le \mu_i \le \bar{y}_i + t_{\alpha/2,n_T-a}\widehat{\sigma}/\sqrt{n_i} \tag{4.25}$$

For the difference in two treatment means, we use Eq. (2.94) which is for the case where the variances are unknown but equal. Then, with $S_p = \widehat{\sigma}$ and notational changes

$$\begin{aligned}\bar{y}_i - \bar{y}_k - t_{\alpha/2,n_T-a}\widehat{\sigma}\sqrt{1/n_i + 1/n_k} \le \mu_i - \mu_k \le \\ \bar{y}_i - \bar{y}_k + t_{\alpha/2,n_T-a}\widehat{\sigma}\sqrt{1/n_i + 1/n_k} \quad i \ne k\end{aligned} \tag{4.26}$$

When the signs of the upper and lower confidence limits determined by Eq. (4.26) are different, one can say that at the 100(1 − α)% confidence level there isn't any difference between μ_i and μ_k.

These results are illustrated with the following example.

Example 4.1

Consider the data from a single factor four-level experiment shown in Table 4.3. We shall perform a one-factor analysis of variance on these data. The results that follow were obtained with interactive graphic IG4-1. The analysis of variance becomes that shown in Table 4.4. It is seen from the p-value in this table that there is a reasonably high probability that at least one treatment mean is different from the other treatment means. Continuing with interactive graphic IG4-1, we obtain the confidence limits on μ_i, which are shown in Table 4.5.

The data shown in Table 4.3 and the tabulated results given in Table 4.5 are displayed as an enhanced box-whisker plot shown in Fig. 4.2. From Fig. 4.2, it is seen that the mean of treatment A_2 is most likely different from the mean of

Table 4.3 Data for Example 4.1

A_1 ($n_1 = 21$)		A_2 ($n_2 = 10$)	A_3 ($n_3 = 10$)	A_4 ($n_4 = 8$)
49.22	105.9	96.44	62.12	110.3
45.54	58.75	74.02	94.07	56.43
45.04	86.37	68.34	142.7	117.6
95.65	59.31	91.12	52.15	78.35
30.99	73.77	105.8	175.9	149.2
35.84	116.2	−0.40	79.71	82.66
81.89	45.68	1.57	29.31	110.8
87.41	70.99	0.67	78.1	92.68
105.1	76.58	0.70	128.3	
95.11	83.64	100.5	132.3	
97.08				

Table 4.4 Analysis of variance for a single factor four level experiment for the data in Table 4.3

Source of variation	Degrees of freedom	Sum of squares	Mean square	F value	p-value
Regression	3	13,773	4591	3.608	0.0203
Error	45	57,253	1272.3		
Total	48	71,026			

Table 4.5 Confidence interval (CI) for μ_i for a single factor four level experiment using the data in Table 4.3

i	n_i	$\bar{y}_i$	s_i	Lower CI μ_i	Upper CI μ_i
1	21	73.62	25.03	57.91	89.33
2	10	53.88	47.17	31.12	76.64
3	10	97.47	45.97	74.71	120.2
4	8	99.76	28.48	74.31	125.2

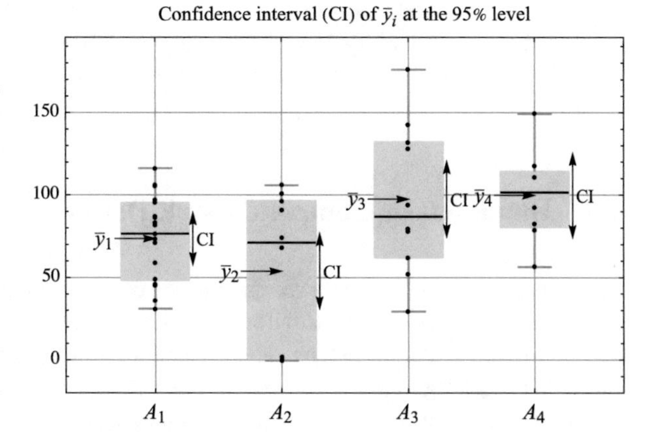

Fig. 4.2 Box-whisker plot of the data in Table 4.3 represented by filled circles combined with the results of Table 4.5

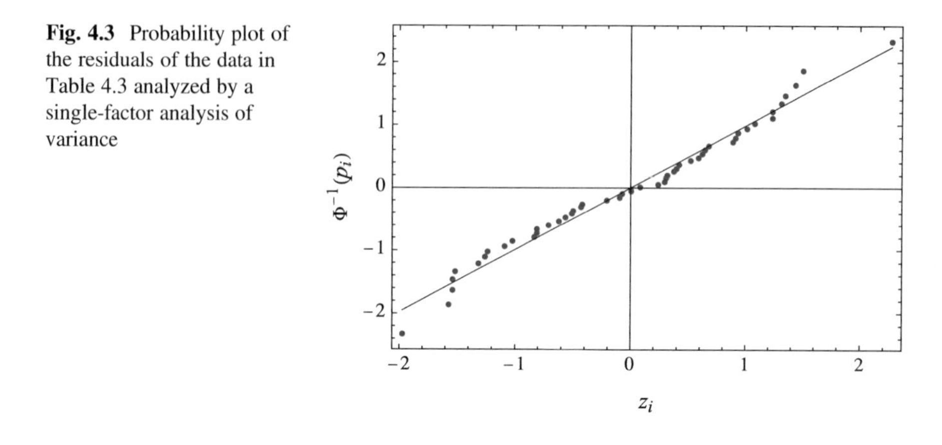

Fig. 4.3 Probability plot of the residuals of the data in Table 4.3 analyzed by a single-factor analysis of variance

treatments A_3 and A_4 since their confidence intervals don't overlap at the 95% confidence level. To verify this conjecture, we use Eq. (4.26) to find that

$$53.88 - 99.76 - 2.0154\sqrt{1272.3}\sqrt{\frac{1}{10}+\frac{1}{8}} \le \mu_2 - \mu_4 \le$$

$$53.88 - 99.76 + 2.0154\sqrt{1272.3}\sqrt{\frac{1}{10}+\frac{1}{8}}$$

$$-79.98 \le \mu_2 - \mu_4 \le -11.78$$

where, from interactive graphic IG2-2, we find that $t_{0.025,44} = 2.0154$. Thus, since the upper and lower confidence limits are both negative, we can say that $\mu_4 > \mu_2$. A similar analysis on the means of treatments A_3 and A_4 yields $-36.39 \le \mu_3 - \mu_4 \le 31.81$. Since the signs of the upper and lower confidence limits are different, we conclude that there is no significant different difference between these means at the 95% confidence level.

Next, we determine the residuals, which are given by $e_{ij} = y_{ij} - \bar{y}_{i\cdot}$. They are displayed in a probability plot shown in Fig. 4.3, where it is seen that they are normally distributed. In obtaining Fig. 4.3, Eq. (2.41) was used.

4.3 Randomized Complete Block Design

Blocking is an extension of paired samples discussed at the end of Sect. 2.6. It is used to eliminate variations due to an extraneous source when a comparison is made between two population means. The extraneous source in the present discussion is called the blocking variable and can be such things as variations due to day-to-day operations, persons involved in the operations, and batch to batch variations in materials and artifacts used in the operations. To employ blocking, one selects the blocking variable and then, with this blocking variable fixed, runs the experiment at all the treatment levels. For example, if day-to-day variations were chosen as the blocking variable, then on a given day all the treatments levels would be run and completed on that day. The order in which each treatment level is run must be chosen randomly and independently for each block. In addition, only one replicate is obtained at each treatment level. The advantage of this procedure is that when treatment comparisons are made, they indicate only variations within the block; block to block variations have been eliminated. It is mentioned that blocking is not a factor, it is simply a different way of performing the experiment and analyzing the results. The objective is still to determine if any of the treatments (factor levels) influence the outcome. Since each block contains all the treatments, the procedure is known as a complete block design.

The complete block design analysis is similar to that performed in Sect. 4.2 for the single factor analysis of variance, but with significant changes in notation and in the model. Referring to the data shown in Table 4.6, at each blocking variable $j = 1, 2, \ldots, b$ we record one response y_{ij}, at each of the $A_1, A_2, \ldots, A_a$ levels. As mentioned previously, the order in which the observations in each B_j column of Table 4.6 were obtained randomly such that the run order in each column is determined independently of the run order for the other columns. The total number of observations is $n_T = ab$. The treatment mean is given by

$$\bar{y}_{i\cdot} = \frac{1}{b}\sum_{j=1}^{b} y_{ij} \tag{4.27}$$

and the blocking mean is given by

Table 4.6 Randomized complete block design, where A_i, $i = 1, 2, \ldots, a$ and B_j, $j = 1, 2, \ldots, b$

	B_1	$\cdots$	B_k	$\cdots$	B_b	Treatment means $\bar{y}_{i\cdot} = \frac{1}{b}\sum_{j=1}^{b} y_{ij}$
A_1	y_{11}		y_{1k}		y_{1b}	$\bar{y}_{1\cdot}$
$\vdots$	$\vdots$		$\vdots$		$\vdots$	$\vdots$
A_m	y_{m1}		y_{mk}		y_{mb}	$\bar{y}_{m\cdot}$
$\vdots$	$\vdots$		$\vdots$		$\vdots$	$\vdots$
A_a	y_{a1}		y_{ak}		y_{ab}	$\bar{y}_{a\cdot}$
Block means $\bar{y}_{\cdot j} = \frac{1}{a}\sum_{i=1}^{a} y_{ij}$	$\bar{y}_{\cdot 1}$		$\bar{y}_{\cdot k}$		$\bar{y}_{\cdot b}$	Grand mean $\bar{y} = \frac{1}{ab}\sum_{i=1}^{a}\sum_{j=1}^{b} y_{ij}$

$$\bar{y}_{\cdot j} = \frac{1}{a}\sum_{i=1}^{a} y_{ij} \tag{4.28}$$

where the '•' in the subscript indicates that the summation has been performed on that index.

We assume a linear model for the response y_{ij} as

$$y_{ij} = \bar{y} + \tau_i + \beta_j + \varepsilon_{ij} \quad i = 1, 2, \ldots, a \quad j = 1, 2, \ldots, b \tag{4.29}$$

where ε_{ij} represents the independent normal random error with each $\varepsilon_{ij} \sim N(0,\sigma^2)$, $\bar{y}$ is the overall (grand) mean given by

$$\bar{y} = \frac{1}{ab}\sum_{i=1}^{a}\sum_{j=1}^{b} y_{ij} = \frac{1}{a}\sum_{i=1}^{a}\bar{y}_{i\cdot} = \frac{1}{b}\sum_{j=1}^{b}\bar{y}_{\cdot j} \tag{4.30}$$

and τ_i and β_j are subject to the following restrictions

$$\sum_{i=1}^{a}\tau_i = 0 \qquad \sum_{j=1}^{b}\beta_j = 0$$

The model given by Eq. (4.29) assumes that there is no interaction between the treatment levels and the blocking variable. This model for a randomized complete block design is sometimes thought of as a balanced two-way analysis of variance without replication or interaction.

We now obtain the *sum of squares identity* to partition the variability. Following the procedure used to derive Eq. (4.4), we find that when a blocking variable is considered the identity is written as

$$\sum_{i=1}^{a}\sum_{j=1}^{b}\left(y_{ij} - \bar{y}\right)^2 = \sum_{i=1}^{a}\sum_{j=1}^{b}\left(\left(\bar{y}_{i\cdot} - \bar{y}\right) + \left(\bar{y}_{\cdot j} - \bar{y}\right) + \left(y_{ij} - \bar{y}_{i\cdot} - \bar{y}_{\cdot j} + \bar{y}\right)\right)^2$$

which upon expansion can be shown to reduce to

$$\begin{aligned}\sum_{i=1}^{a}\sum_{j=1}^{b}\left(y_{ij} - \bar{y}\right)^2 = b\sum_{i=1}^{a}\left(\bar{y}_{i\cdot} - \bar{y}\right)^2 + a\sum_{j=1}^{b}\left(\bar{y}_{\cdot j} - \bar{y}\right)^2 \\ + \sum_{i=1}^{a}\sum_{j=1}^{b}\left(y_{ij} - \bar{y}_{i\cdot} - \bar{y}_{\cdot j} + \bar{y}\right)^2\end{aligned} \tag{4.31}$$

We write Eq. (4.31) as

$$SS_T = SS_A + SS_B + SS_E \tag{4.32}$$

where

$$\begin{aligned}
SS_T &= \sum_{i=1}^{a}\sum_{j=1}^{b}\left(y_{ij} - \bar{y}\right)^2 = \sum_{i=1}^{a}\sum_{j=1}^{b} y_{ij}^2 - ab\bar{y}^2 && \text{(total sum of squares)} \\
SS_A &= b\sum_{i=1}^{a}\left(\bar{y}_{i\cdot} - \bar{y}\right)^2 = b\sum_{i=1}^{a}\bar{y}_{i\cdot}^2 - ab\bar{y}^2 && \text{(treatment sum of squares)} \\
SS_B &= a\sum_{j=1}^{b}\left(\bar{y}_{\cdot j} - \bar{y}\right)^2 = a\sum_{j=1}^{b}\bar{y}_{\cdot j}^2 - ab\bar{y}^2 && \text{(block sum of squares)} \\
SS_E &= \sum_{i=1}^{a}\sum_{j=1}^{b}\left(y_{ij} - \bar{y}_{i\cdot} - \bar{y}_{\cdot j} + \bar{y}\right)^2 \\
&= \sum_{i=1}^{a}\sum_{j=1}^{b} y_{ij}^2 - b\sum_{i=1}^{a}\bar{y}_{i\cdot}^2 - a\sum_{j=1}^{b}\bar{y}_{\cdot j}^2 + ab\bar{y}^2 && \text{(error sum of squares)}
\end{aligned} \tag{4.33}$$

From the definition of an unbiased variance, we see that SS_A has $a - 1$ degrees of freedom, SS_B has $b - 1$ degrees of freedom, SS_T has $ab - 1$ degrees of freedom, and therefore, SS_E has $(a - 1)(b - 1)$ degrees of freedom.

The mean square values for the treatment, blocking, and error, respectively, are obtained from

$$\text{MSA} = \frac{SS_A}{a-1}, \quad \text{MSB} = \frac{SS_B}{b-1}, \quad \text{MSE} = \frac{SS_E}{(a-1)(b-1)} \tag{4.34}$$

These results are summarized in Table 4.7.

Table 4.7 Analysis of variance of a single factor experiment with blocking, where $a_0 = (a - 1)(b - 1)$

Source of variation	Degrees of freedom	Sum of squares	Mean square	F value	p-value[a]
Treatment	$a - 1$	SS_A	$\text{MSA} = \frac{SS_A}{a-1}$	$F_A = \frac{\text{MSA}}{\text{MSE}}$	$\Psi_f(F_A, a-1, a_0)$
Block	$b - 1$	SS_B	$\text{MSB} = \frac{SS_B}{b-1}$	$F_B = \frac{\text{MSB}}{\text{MSE}}$	$\Psi_f(F_B, b-1, a_0)$
Error	a_0	SS_E	$\text{MSE} = \frac{SS_E}{a_0}$		
Total	$ab - 1$	SS_T			

[a] Ψ_f is determined from Eq. (2.120)

Using the method to arrive at Eq. (4.16), the unbiased estimate of the variance is determined from

$$\widehat{\sigma}^2 = \frac{SS_E}{(a-1)(b-1)} = \text{MSE} \tag{4.35}$$

and using the method to arrive at Eq. (4.22) is it found that

$$\begin{aligned} E\left(\frac{SS_A}{a-1}\right) &= \sigma^2 + \frac{b}{a-1}\sum_{i=1}^{a}\tau_i^2 \\ E\left(\frac{SS_B}{b-1}\right) &= \sigma^2 + \frac{a}{b-1}\sum_{i=1}^{b}\beta_i^2 \end{aligned} \tag{4.36}$$

It is noted that Eq. (4.35) is an unbiased estimate of the variance irrespective of the values of either τ_i or β_j because of the model's assumption that there is no interaction between the treatments and the blocks.

The above results are used to determine the effects of the treatments from examination of the null hypothesis

$$H_0 : \tau_1 = \cdots = \tau_a = 0 \tag{4.37}$$

with the alternative H_A: $\bar{y}_{i\cdot}$ are not all equal. This is the primary objective of the analysis. The corresponding p-value $= \Psi_f(F_A, a-1,(a_0)$, where $a_0 = (a-1)(b-1)$ and

$$F_A = \frac{\text{MSA}}{\text{MSE}} \tag{4.38}$$

The effects of blocking are determined from the examination of the null hypothesis

$$H_0 : \beta_1 = \cdots = \beta_b = 0 \tag{4.39}$$

with the alternative H_A: $\bar{y}_{\cdot j}$ are not all equal. The p-value $= \Psi_f(F_B, b-1, a_0)$, where

$$F_B = \frac{\text{MSB}}{\text{MSE}} \tag{4.40}$$

A very small blocking p-value indicates that there is variability from at least one block.

These results are illustrated with the following example.

Table 4.8 Data from a randomized complete block design and some intermediate calculations for four treatment levels and six blocking levels

↓Treatments Blocks→	B_1	B_2	B_3	B_4	B_5	B_6	$\bar{y}_{i\cdot}$
A_1	41.9	38.7	39.	39.3	42.3	43.0	40.70
A_2	40.2	40.5	40.9	42.7	42.9	43.5	41.78
A_3	40.4	40.7	41.5	43.6	45.1	45.3	42.77
A_4	41.0	41.9	43.2	43.9	45.6	42.0	42.93
$\bar{y}_{\cdot j}$	40.88	40.45	41.15	42.38	43.98	43.45	$\bar{y} = 42.05$

Table 4.9 Analysis of variance for a single factor randomized complete block design for the data in Table 4.8

Source of variation	Degrees of freedom	Sum of squares	Mean square	F value	p-value
Regression	3	19.125	6.3749	4.01	0.0279
Block	5	42.087	8.4174	5.294	0.00533
Error	15	23.848	1.5899		
Total	23	85.06			

Example 4.2
Consider the data from a single factor four-level experiment shown in Table 4.8. The test was run as a randomized complete block design with six different instances of the blocking variable. Thus, $a = 4$ and $b = 6$. Using the interactive graphic IG4-2, we obtain the analysis of variance given in Table 4.9.

It is seen from Table 4.9 that at least one treatment level is different from the other treatment levels at around the 97% confidence level. In addition, the small p-value of the blocking variable indicates that there is block-to-block variability from at least one blocking instance.

4.4 Two Factor Experiment

A two-factor experiment is used to determine the effects of two factors A and B. The factor A has a levels $A_1, A_2, \ldots, A_a$ and the factor B has b levels $B_1, B_2, \ldots, B_b$. In addition, we consider n ($n > 1$) replications (observations) at each of the factor combinations, which makes it a balanced design. The observation is denoted y_{ijk}, where $i = 1, 2, \ldots, a$, $j = 1, 2, \ldots, b$, and $k = 1, 2, \ldots, n$. Hence, the total number of observations is $n_T = abn$. In Table 4.10, the observations have been organized as shown and the definitions of several different means are given. As in the previous experimental designs, the observations were obtained in random order and not in the

Table 4.10 Data layout for a two-factor factorial design and their mean values, where A_i, $i = 1, 2, \ldots, a$ and B_j, $j = 1, 2, \ldots, b$

	B_1	B_2	$\cdots$	B_b	Factor A means $\bar{y}_{i\cdot\cdot} = \frac{1}{bn}\sum_{j=1}^{b}\sum_{k=1}^{n} y_{ijk} = \frac{1}{b}\sum_{j=1}^{b}\bar{y}_{ij\cdot}$
A_1	$y_{111}, y_{112}, \ldots, y_{11n}$ $\bar{y}_{11\cdot} = \frac{1}{n}\sum_{k=1}^{n} y_{11k}$	$y_{121}, y_{122}, \ldots, y_{12n}$ $\bar{y}_{12\cdot} = \frac{1}{n}\sum_{k=1}^{n} y_{12k}$	$\cdots$	$y_{1b1}, y_{1b2}, \ldots, y_{1bn}$ $\bar{y}_{1b\cdot} = \frac{1}{n}\sum_{k=1}^{n} y_{1bk}$	$\bar{y}_{1\cdot\cdot}$
A_2	$y_{211}, y_{212}, \ldots, y_{21n}$ $\bar{y}_{21\cdot} = \frac{1}{n}\sum_{k=1}^{n} y_{21k}$	$y_{221}, y_{222}, \ldots, y_{22n}$ $\bar{y}_{22\cdot} = \frac{1}{n}\sum_{k=1}^{n} y_{22k}$	$\cdots$	$y_{2b1}, y_{2b2}, \ldots, y_{2bn}$ $\bar{y}_{2b\cdot} = \frac{1}{n}\sum_{k=1}^{n} y_{2bk}$	$\bar{y}_{2\cdot\cdot}$
$\vdots$	$\vdots$	$\vdots$		$\vdots$	$\vdots$
A_a	$y_{a11}, y_{a12}, \ldots, y_{a1n}$ $\bar{y}_{a1\cdot} = \frac{1}{n}\sum_{k=1}^{n} y_{a1k}$	$y_{a21}, y_{a22}, \ldots, y_{a2n}$ $\bar{y}_{a2\cdot} = \frac{1}{n}\sum_{k=1}^{n} y_{a2k}$	$\cdots$	$y_{ab1}, y_{ab2}, \ldots, y_{abn}$ $\bar{y}_{ab\cdot} = \frac{1}{n}\sum_{k=1}^{n} y_{abk}$	$\bar{y}_{a\cdot\cdot}$
Factor B means $\bar{y}_{\cdot j\cdot} = \frac{1}{an}\sum_{i=1}^{a}\sum_{k=1}^{n} y_{ijk} = \frac{1}{a}\sum_{i=1}^{a}\bar{y}_{ij\cdot}$	$\bar{y}_{\cdot 1\cdot}$	$\bar{y}_{\cdot 2\cdot}$	$\cdots$	$\bar{y}_{\cdot b\cdot}$	Grand mean $\bar{y} = \frac{1}{abn}\sum_{i=1}^{a}\sum_{j=1}^{b}\sum_{k=1}^{n} y_{ijk} = \frac{1}{ab}\sum_{i=1}^{a}\sum_{j=1}^{b}\bar{y}_{ij\cdot}$

order that one may infer from its position in the table. That is, the n_T runs are assigned a number from 1 to n_T and then these numbers are ordered randomly.

The mean of factor A is

$$\bar{y}_{i\cdot\cdot} = \frac{1}{bn} \sum_{j=1}^{b} \sum_{k=1}^{n} y_{ijk} \tag{4.41}$$

The mean of factor B is

$$\bar{y}_{\cdot j\cdot} = \frac{1}{an} \sum_{i=1}^{a} \sum_{k=1}^{n} y_{ijk} \tag{4.42}$$

and the overall (grand) mean is

$$\bar{y} = \frac{1}{abn} \sum_{i=1}^{a} \sum_{j=1}^{b} \sum_{k=1}^{n} y_{ijk} = \frac{1}{ab} \sum_{i=1}^{a} \sum_{j=1}^{b} \bar{y}_{ij\cdot} \tag{4.43}$$

The quantity $\bar{y}_{ij\cdot}$ is called ijth cell's average observation.

We assume a linear model for the response y_{ijk} as

$$y_{ijk} = \bar{y} + \tau_i + \beta_j + (\tau\beta)_{ij} + \varepsilon_{ijk} \tag{4.44}$$

where $i = 1, 2, \ldots, a$, $j = 1, 2, \ldots, b$, $k = 1, 2, \ldots, n$, τ_i is the effect of the ith level of factor A, β_j is the effect of the jth level of factor B, $(\tau\beta)_{ij}$ is the effect of the interaction between A and B, and $\varepsilon_{ijk} \sim N(0,\sigma^2)$ is the random error. The inclusion of an interaction effect between the two factors is a way to determine if the two factors operate independently. The quantities τ_i, β_j, and $(\tau\beta)_{ij}$ are subject to the following restrictions

$$\sum_{i=1}^{a} \tau_i = 0 \qquad \sum_{j=1}^{b} \beta_j = 0 \qquad \sum_{i=1}^{a} (\tau\beta)_{ij} = 0 \qquad \sum_{j=1}^{b} (\tau\beta)_{ij} = 0$$

The sum of squares identity for the two-factor experiments can be shown to be

$$\sum_{i=1}^{a}\sum_{j=1}^{b}\sum_{k=1}^{n}\left(y_{ijk}-\overline{y}\right)^2 = bn\sum_{i=1}^{a}\left(\overline{y}_{i\cdot\cdot}-\overline{y}\right)^2 + an\sum_{j=1}^{b}\left(\overline{y}_{\cdot j\cdot}-\overline{y}\right)^2 + \sum_{i=1}^{a}\sum_{j=1}^{b}\left(\overline{y}_{ij\cdot}-\overline{y}_{i\cdot\cdot}-\overline{y}_{\cdot j\cdot}+\overline{y}\right)^2 + \sum_{i=1}^{a}\sum_{j=1}^{b}\sum_{k=1}^{n}\left(y_{ijk}-\overline{y}_{ij\cdot}\right)^2 \tag{4.45}$$

Equation (4.45) can be written as

$$SS_T = SS_A + SS_B + SS_{AB} + SS_E \tag{4.46}$$

where

$$\begin{aligned}
SS_T &= \sum_{i=1}^{a}\sum_{j=1}^{b}\sum_{k=1}^{n}\left(y_{ijk}-\overline{y}\right)^2 = \sum_{i=1}^{a}\sum_{j=1}^{b}\sum_{k=1}^{n} y_{ijk}^2 - abn\overline{y}^2 && \text{(total sum of squares)}\\
SS_A &= bn\sum_{i=1}^{a}\left(\overline{y}_{i\cdot\cdot}-\overline{y}\right)^2 = bn\sum_{i=1}^{a}\overline{y}_{i\cdot\cdot}^2 - abn\overline{y}^2 && \text{(factor } A \text{ sum of squares)}\\
SS_B &= an\sum_{j=1}^{b}\left(\overline{y}_{\cdot j\cdot}-\overline{y}\right)^2 = an\sum_{j=1}^{b}\overline{y}_{\cdot j\cdot}^2 - abn\overline{y}^2 && \text{(factor } B \text{ sum of squares)}\\
SS_{AB} &= n\sum_{i=1}^{a}\sum_{j=1}^{b}\left(\overline{y}_{ij\cdot}-\overline{y}_{i\cdot\cdot}-\overline{y}_{\cdot j\cdot}+\overline{y}\right)^2 && \text{(interaction sum of squares)}\\
SS_E &= \sum_{i=1}^{a}\sum_{j=1}^{b}\sum_{k=1}^{n}\left(y_{ijk}-\overline{y}_{ij\cdot}\right)^2\\
&= \sum_{i=1}^{a}\sum_{j=1}^{b}\sum_{k=1}^{n} y_{ijk}^2 - n\sum_{i=1}^{a}\sum_{j=1}^{b}\overline{y}_{ij\cdot}^2 && \text{(error sum of squares)}
\end{aligned} \tag{4.47}$$

The mean square values for the factors, interactions, and error, respectively, are

$$\begin{aligned}
\text{MSA} &= \frac{SS_A}{a-1}, \quad \text{MSB} = \frac{SS_B}{b-1},\\
\text{MSAB} &= \frac{SS_{AB}}{(a-1)(b-1)}, \quad \text{MSE} = \frac{SS_E}{ab(n-1)}
\end{aligned} \tag{4.48}$$

Using the method to arrive at Eq. (4.16), the estimate of the variance is determined from

$$E\left(\frac{SS_E}{ab(n-1)}\right) = E(\text{MSE}) = \sigma^2$$

and, therefore, an unbiased estimator of the error variance is

$$\widehat{\sigma}^2 = \frac{SS_E}{ab(n-1)} = \text{MSE} \tag{4.49}$$

Using the method to arrive at Eq. (4.22), it is found that

$$\begin{aligned} E\left(\frac{SS_A}{a-1}\right) &= \sigma^2 + \frac{bn}{a-1}\sum_{i=1}^{a}\tau_i^2 \\ E\left(\frac{SS_B}{b-1}\right) &= \sigma^2 + \frac{an}{b-1}\sum_{i=1}^{b}\beta_i^2 \\ E\left(\frac{SS_{AB}}{(a-1)(b-1)}\right) &= \sigma^2 + \frac{n}{(a-1)(b-1)}\sum_{i=1}^{a}\sum_{j=1}^{b}(\tau\beta)_{ij}^2 \end{aligned} \tag{4.50}$$

These results are used to determine the effects of the factors and interactions from examination of the following null hypotheses. Starting with the interaction AB, the null hypothesis is

$$H_0 : (\tau\beta)_{ij} = 0 \quad i = 1, 2, \ldots a \quad j = 1, 2, \ldots b \tag{4.51}$$

with the alternative H_A: $(\tau\beta)_{ij}$ are not all equal. The p-value $= \Psi_f(F_{AB}, b-1, ab(n-1))$, where $F_{AB} =$ MSAB/MSE. If the null hypothesis is rejected, then the presence of an interaction effect has been established. However, when there is interaction, the factors A and B may not have any useful interpretive value since the factors are not independent of each other. In fact, knowing if the factors are interacting is sometimes of more value than knowledge about the factors themselves. Regarding the factor A, the null hypothesis is

$$H_0 : \tau_1 = \cdots = \tau_a = 0 \tag{4.52}$$

with the alternative H_A: $\bar{y}_{i.}$ are not all equal. The p-value $= \Psi_f(F_{AB}, a-1, ab(n-1))$, where $F_A =$ MSA/MSE. For the factor B, the null hypothesis is

$$H_0 : \beta_1 = \cdots = \beta_b = 0 \tag{4.53}$$

with the alternative H_A: $\bar{y}_{.j}$ are not all equal. The p-value $= \Psi_f(F_{AB}, b-1, ab(n-1))$, where $F_B =$ MSB/MSE.

These results are summarized in Table 4.11.

Table 4.11 Two-factor analysis of variance, where $a_0 = a - 1$, $b_0 = b - 1$, and $c_0 = ab(n - 1)$

Source of variation	Degrees of freedom	Sum of squares	Mean square	*F* value	*p*-value[a]
Factor *A*	a_0	SS_A	$\text{MSA} = \dfrac{SS_A}{a_0}$	$F_A = \dfrac{\text{MSA}}{\text{MSE}}$	$\Psi_f(F_A, a_0, c_0)$
Factor *B*	b_0	SS_B	$\text{MSB} = \dfrac{SS_B}{b_0}$	$F_B = \dfrac{\text{MSB}}{\text{MSE}}$	$\Psi_f(F_B, b_0, c_0)$
Interaction *AB*	$a_0 b_0$	SS_{AB}	$\text{MSAB} = \dfrac{SS_{AB}}{a_0 b_0}$	$F_{AB} = \dfrac{\text{MSAB}}{\text{MSE}}$	$\Psi_f(F_{AB}, a_0 b_0, c_0)$
Error	c_0	SS_E	$\text{MSE} = \dfrac{SS_E}{c_0}$		
Total	$abn - 1$	SS_T			

[a] Ψ_f is determined from Eq. (2.120)

The model's adequacy is determined with a probability plot of the residuals, which are defined as the difference between the observations and the cell's average; that is,

$$e_{ijk} = y_{ijk} - \bar{y}_{ij\cdot} \tag{4.54}$$

These results are illustrated with the following example.

Example 4.3
Consider the data from a two-factor experiment shown in Table 4.12, which includes the determination of their various means. Using interactive graphic IG4-3, the analysis of variance given by Table 4.13 is obtained. From this table, it is seen from the magnitude of the *p*-value that there is no interaction between factors *A* and *B*. This can be verified by plotting the averages $\bar{y}_{ij\cdot}$ as shown in Fig. 4.4, which is also obtained from interactive graphic IG4-3. It is seen that the shapes of the connected means are approximately the same when one shape is shifted such that the values at A_2 are coincident; in this instance it can be visualized that the two sets of lines are nearly coincident. Coincidence indicates the absence of an interaction. Since the factors are uncoupled, one can make a judgment on the average values as to which levels are better depending on whether one is interested in $\bar{y}_{ij\cdot}$ being small or large. The small *p*-value for the *B*-factor indicates that the two levels are different, as is seen in Fig. 4.4. Based on the *p*-value for the *A* factor, there is some evidence that at least one level is different from the other levels. This is seen in Fig. 4.4.

The model adequacy is determined from a probability plot of the residuals given by Eq. (4.54). Using the procedure to arrive at Fig. 2.7, we obtain the results shown in Fig. 4.5, where it is seen that the data are normally distributed. In obtaining Fig. 4.5, Eq. (2.41) was used. This result is also obtained from interactive graphic IG4-3.

Table 4.12 Data for the two-factor experiment of Example 4.3 and their various means

Factor A Levels	Factor B Levels: B_1	Factor B Levels: B_2	Factor A means
A_1	4.8, 5.4, 5.16 $\bar{y}_{11\cdot} = 5.12$	6.48, 5.88, 6.72 $\bar{y}_{12\cdot} = 6.36$	$\bar{y}_{1\cdot\cdot} = 5.74$
A_2	5.6, 4.9, 5.4 $\bar{y}_{21\cdot} = 5.3$	6.38, 6.71, 6.93 $\bar{y}_{22\cdot} = 6.67$	$\bar{y}_{2\cdot\cdot} = 5.99$
A_3	4.56, 4.44, 4.8 $\bar{y}_{31\cdot} = 4.6$	6.6, 6.0, 6.0 $\bar{y}_{32\cdot} = 6.2$	$\bar{y}_{3\cdot\cdot} = 5.4$
Factor B means	$\bar{y}_{\cdot 1\cdot} = 5.007$	$\bar{y}_{\cdot 2\cdot} = 6.411$	$\bar{y} = 5.709$

Table 4.13 Analysis of variance for the data in Table 4.12

Source of variation	Degrees of freedom	Sum of squares	Mean square	F value	p-value
Factor A	2	1.041	0.5206	4.891	0.02795
Factor B	1	8.876	8.876	83.39	9.47×10^{-7}
Interaction AB	2	0.0994	0.0497	0.467	0.638
Error	12	1.277	0.1064		
Total	17	11.29			

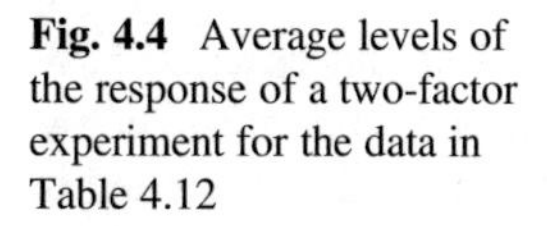
Fig. 4.4 Average levels of the response of a two-factor experiment for the data in Table 4.12

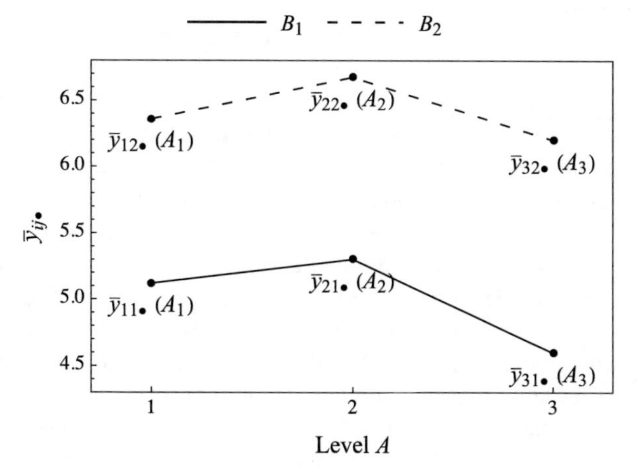

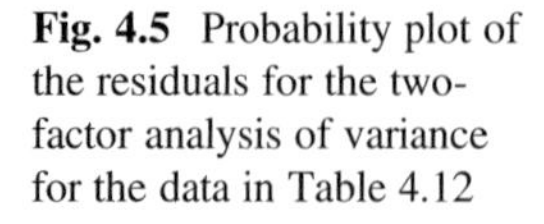
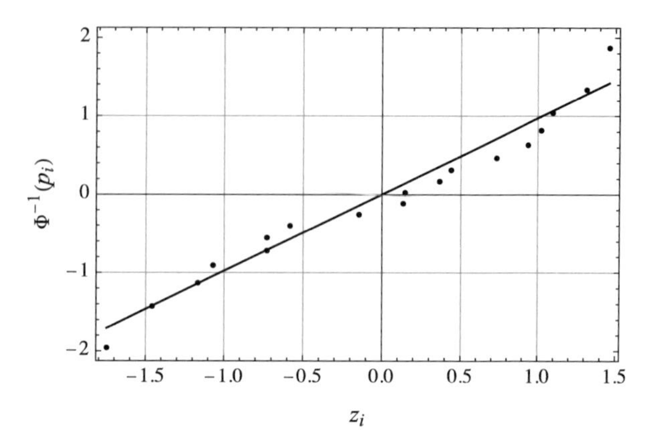

Fig. 4.5 Probability plot of the residuals for the two-factor analysis of variance for the data in Table 4.12

4.5 2^k-Factorial Experiments

A 2^k factorial experiment is an experiment in which there are k factors with each factor being examined at two levels. A complete factorial experiment requires that every combination of each level of each factor is run thereby resulting in 2^k runs. If there are n ($n > 1$) replicates at each combination, then a total of $n2^k$ runs are made. Frequently, the level with the larger numerical value is denoted the high level and the one with the smaller numerical value the low level. Since only two levels are being used, a linear relationship over the range of the two levels is implied. The 2^k factorial experiment provides the ability to find the main effects and any of their interactions and is often used as a screening experiment, one that determines which factors are important and the appropriate ranges of the important factors. Then, from an analysis of the results of the screening experiment one or more follow up experiments can be designed.

The simplest 2^k factorial experiment is that for which $k = 2$. In this case, we have a total of four (2^2) treatment combinations, with each combination run n times, where $n > 1$. This case is a special case of the two-factor experiment discussed in Sect. 4.4. It is re-examined with the goal of extending it to a 2^k factorial design. Thus, to discuss the 2^2-factorial experiment, we introduce the notation shown in Table 4.14. When both factors are at their low level the sum of their replicates is denoted (1)—parentheses required. When factor A is at its high level and factor B is at its low level the sum of the replicates is denoted by a lower-case letter of that uppercase letter used to denote the high-level factor. Thus, a indicates the sum of the replicates when A is at its high level and B is at its low level. Since this is a completely randomized design, each observation in Table 4.14 is obtained in a randomly determined order and not in the order that one may infer from its position in the table. That is, the $n2^2$ runs are assigned a number from 1 to $n2^2$ and then these numbers are ordered randomly. This ordering process is also used for 2^k-factor experiments, $k > 2$.

Table 4.14 Notation for a 2^2-factorial experiment

Factor A	Factor B: $B_1 = B_{\text{Low}}$	Factor B: $B_2 = B_{\text{High}}$	Factor A Main effects average
$A_1 = A_{\text{Low}}$	$y_{111}, y_{112}, \ldots, y_{11n}$ $(1) = \sum_{m=1}^{n} y_{11m}$ $A_1B_1 \Rightarrow (1)(1) = (1)$	$y_{121}, y_{122}, \ldots, y_{12n}$ $b = \sum_{m=1}^{n} y_{12m}$ $A_1B_2 \Rightarrow (1)b = b$	$y_{1b} = \frac{1}{2n}((1) + b)$
$A_2 = A_{\text{High}}$	$y_{211}, y_{212}, \ldots, y_{21n}$ $a = \sum_{m=1}^{n} y_{21m}$ $A_2B_1 \Rightarrow a(1) = a$	$y_{221}, y_{222}, \ldots, y_{22n}$ $ab = \sum_{m=1}^{n} y_{22m}$ $A_2B_2 \Rightarrow ab$	$y_{2b} = \frac{1}{2n}(a + ab)$
Factor B Main effects average	$y_{a1} = \frac{1}{2n}((1) + a)$	$y_{a2} = \frac{1}{2n}(b + ab)$	

The main effect of factor A is given by

$$\text{Effect}_A = y_{2b} - y_{1b} = \frac{1}{2n}(a + ab - (1) - b) \tag{4.55}$$

and that for the main effect of factor B is

$$\text{Effect}_B = y_{a2} - y_{a1} = \frac{1}{2n}(b + ab - (1) - a) \tag{4.56}$$

The interaction of the factors is obtained by taking the difference of the average values of the diagonal terms in the cells in Table 4.14; that is,

$$\begin{aligned}\text{Effect}_{AB} &= \frac{1}{2n}((1) + ab) - \frac{1}{2n}(a + b) \\ &= \frac{1}{2n}((1) + ab - a - b)\end{aligned} \tag{4.57}$$

The above quantities and those in Table 4.14 are given a graphical interpretation in Fig. 4.6. From Eq. (4.57) and Fig. 4.6, we see that when the lines are parallel $b - (1) = ab - a$ and, consequently, $\text{Effect}_{AB} = 0$. Therefore, parallel (or, in a practical sense, almost parallel) lines in a plot of the average responses indicate that the factors A and B do not interact.

We now introduce a quantity called a contrast, denoted C_γ, which are the terms appearing inside the parentheses in Eqs. (4.55)–(4.57); that is,

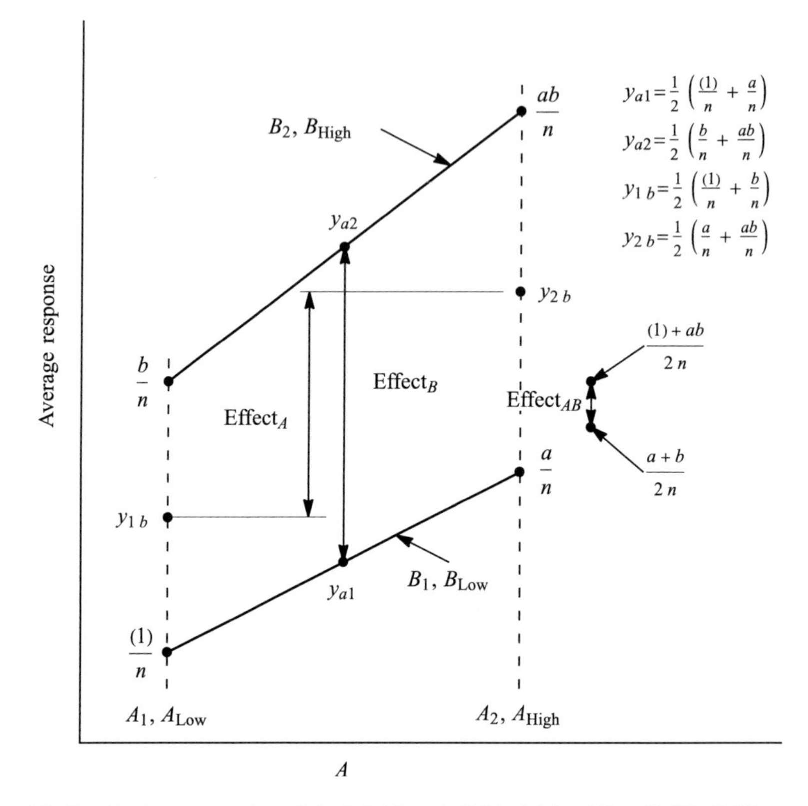

Fig. 4.6 Graphical representation of the definitions in Table 4.14 and Eqs. (4.55)–(4.57)

$$\begin{aligned} C_A &= -(1) + a - b + ab \\ C_B &= -(1) - a + b + ab \\ C_{AB} &= (1) - a - b + ab \end{aligned} \tag{4.58}$$

Notice that all three contrasts contain the same terms, but with the coefficients of each term in the contrasts being either a +1 or a −1. To see the utility of this observation, we place coefficients of each term in the contrasts in Eq. (4.58) in tabular form as shown in Table 4.15, where, by convention, the '1' is omitted. Therefore, Eqs. (4.55)–(4.57) can be written as

$$\text{Effect}_\gamma = \frac{1}{n2^1} C_\gamma \quad \gamma = A, B, AB \tag{4.59}$$

From Table 4.15, it is seen that the sign of the interaction *AB* is the product of the respective signs of the coefficients of contrasts *A* and *B*.

To determine the sum of squares of factor *A*, we find that for a 2^2-factorial experiment that Eq. (4.47) can be written as

Table 4.15 Signs of the terms appearing in the Contrasts C_γ for a 2^2-factorial experiment

Treatment combination	γ		
	A	B	AB
(1)	−	−	+
a	+	−	−
b	−	+	−
ab	+	+	+

$$SS_A = 2n\left[(\bar{y}_{1\cdot\cdot} - \bar{y})^2 + (\bar{y}_{2\cdot\cdot} - \bar{y})^2\right] \tag{4.60}$$

However, referring to Tables 4.10 and 4.14, we note that

$$\bar{y}_{11\cdot} = \frac{1}{n}(1), \quad \bar{y}_{12\cdot} = \frac{1}{n}b, \quad \bar{y}_{21\cdot} = \frac{1}{n}a, \quad \bar{y}_{22\cdot} = \frac{1}{n}ab$$

Therefore,

$$\begin{aligned}\bar{y}_{1\cdot\cdot} &= \frac{1}{2}(\bar{y}_{11\cdot} + \bar{y}_{12\cdot}) = \frac{1}{2n}((1) + b) \\ \bar{y}_{2\cdot\cdot} &= \frac{1}{2}(\bar{y}_{21\cdot} + \bar{y}_{22\cdot}) = \frac{1}{2n}(a + ab)\end{aligned} \tag{4.61}$$

and

$$\bar{y} = \frac{1}{2}(\bar{y}_{1\cdot\cdot} + \bar{y}_{2\cdot\cdot}) = \frac{1}{n2^2}C_I = \frac{1}{2}\text{Effect}_I \tag{4.62}$$

where

$$\begin{aligned}C_I &= (1) + a + b + ab \\ \text{Effect}_I &= \frac{1}{n2^1}C_I\end{aligned} \tag{4.63}$$

Upon substituting Eqs. (4.61) and (4.62) into Eq. (4.60), we arrive at

$$SS_A = \frac{1}{n2^2}(-(1) + a - b + ab)^2 = \frac{1}{n2^2}C_A^2 \tag{4.64}$$

In a similar manner, we find that

$$SS_B = \frac{1}{n2^2} C_B^2$$
$$SS_{AB} = \frac{1}{n2^2} C_{AB}^2 \tag{4.65}$$

Each of the sum of squares in Eqs. (4.64) and (4.65) has one degree of freedom. Hence, the mean square response is simply

$$MS_\gamma = SS_\gamma \quad \gamma = A, B, AB \tag{4.66}$$

The total sum of squares SS_T is computed by using Eq. (4.47), which in the present case becomes

$$SS_T = \sum_{i=1}^{2} \sum_{j=1}^{2} \sum_{m=1}^{n} y_{ijm}^2 - 2^2 n \bar{y}^2 \tag{4.67}$$

where $\bar{y}$ is given by Eq. (4.62). It is noted from Table 4.14 that the subscript $i = 1$ denotes that factor A is at its low level and the subscript $i = 2$ denotes that factor A is at its high level. Similarly, for the subscript j and factor B. From Table 4.8, it is found that SS_T has $2^2 n - 1$ degrees of freedom. The error sum of squares is obtained from Eq. (4.46) as

$$SS_E = SS_T - SS_A - SS_B - SS_{AB} \tag{4.68}$$

which has $2^2(n - 1)$ degrees of freedom. The preceding results are placed in Table 4.16.

Table 4.16 2^2-factorial analysis of variance, where $c_0 = 2^2(n - 1)$

Source of variation	Degrees of freedom	Sum of squares	Mean square	F value	p-value[a]
Factor A	1	SS_A	$MS_A = SS_A$	$F_A = \frac{MS_A}{MS_E}$	$\Psi_f(F_A, 1\ c_0)$
Factor B	1	SS_B	$MS_B = SS_B$	$F_B = \frac{MS_B}{MS_E}$	$\Psi_f(F_B, 1, c_0)$
Interaction AB	1	SS_{AB}	$MS_{AB} = SS_{AB}$	$F_{AB} = \frac{MS_{AB}}{MS_E}$	$\Psi_f(F_{AB}, 1, c_0)$
Error	c_0	SS_E	$MS_E = \frac{SS_E}{c_0}$		
Total	$2^2 n - 1$	SS_T			

[a] Ψ_f is determined from Eq. (2.120)

It is noted that the contrasts given by Eq. (4.58) can be obtained by 'converting' Table 4.12 to matrix form. This conversion will make it easier to generalize these results for $k \geq 3$. Let

$$\boldsymbol{C} = \begin{Bmatrix} C_A \\ C_B \\ C_{AB} \end{Bmatrix} \quad \boldsymbol{g} = \begin{Bmatrix} (1) \\ a \\ b \\ ab \end{Bmatrix} \quad \boldsymbol{h} = \begin{pmatrix} -1 & -1 & 1 \\ 1 & -1 & -1 \\ -1 & 1 & -1 \\ 1 & 1 & 1 \end{pmatrix} \tag{4.69}$$

Then

$$\boldsymbol{C} = \boldsymbol{h}'\boldsymbol{g} \tag{4.70}$$

where, as before, the prime (′) denotes the transpose of the matrix.

We now view these results another way so that we shall be able to use multiple regression analysis to fit a line or surface and be able to compute the residuals. We first introduce a quantity called the coded value c_j, $j = 1, 2$, which corresponds to an input factor level L_j, $j = 1, 2$, where $L_2 > L_1$. Let

$$\Delta = \frac{1}{2}(L_2 - L_1) \qquad L_m = \frac{1}{2}(L_2 + L_1)$$

Then the coded value is

$$c_j = \frac{L_j - L_m}{\Delta} = \frac{2L_j - (L_2 + L_1)}{L_2 - L_1} \qquad j = 1, 2$$

from which it is seen that

$$c_1 = \frac{2L_1 - (L_2 + L_1)}{L_2 - L_1} = -1$$
$$c_2 = \frac{2L_2 - (L_2 + L_1)}{L_2 - L_1} = +1$$

For factor A, $L_2 = A_{\text{High}}$ and $L_1 = A_{\text{Low}}$ and for factor B, $L_2 = B_{\text{High}}$ and $L_1 = B_{\text{Low}}$.

We assume a multiple regression model as

$$\widehat{y} = \widehat{\beta}_0 + \widehat{\beta}_1 x_1 + \widehat{\beta}_2 x_2 + \widehat{\beta}_3 x_1 x_2 \tag{4.71}$$

where x_1 is the coded value for the A levels, x_2 is the coded value for the B levels, and the product $x_1 x_2$ represents the interaction AB.

To determine the values for the estimates of $\widehat{\beta}_j$, $j = 0, 1, 2, 3$, we follow Eq. (3.58) and define the following matrix $\boldsymbol{X}$ and column vectors $\widehat{\boldsymbol{\beta}}$ and $\boldsymbol{y}$ as

$$\boldsymbol{X} = \begin{pmatrix} 1 & -1 & -1 & 1 \\ 1 & 1 & -1 & -1 \\ 1 & -1 & 1 & -1 \\ 1 & 1 & 1 & 1 \end{pmatrix} \quad \widehat{\boldsymbol{\beta}} = \begin{Bmatrix} \widehat{\beta}_0 \\ \widehat{\beta}_1 \\ \widehat{\beta}_2 \\ \widehat{\beta}_3 \end{Bmatrix} \quad \boldsymbol{y} = \frac{1}{n}\begin{Bmatrix} (1) \\ a \\ b \\ ab \end{Bmatrix} \tag{4.72}$$

Then, from Eqs. (3.64), (4.58), (4.63), and (4.72), we have that

$$\begin{aligned} \widehat{\boldsymbol{\beta}} = \begin{Bmatrix} \widehat{\beta}_0 \\ \widehat{\beta}_1 \\ \widehat{\beta}_2 \\ \widehat{\beta}_3 \end{Bmatrix} &= (\boldsymbol{X}'\boldsymbol{X})^{-1}\boldsymbol{X}'\boldsymbol{y} = \frac{1}{n2^2}\begin{pmatrix} 1 & 1 & 1 & 1 \\ -1 & 1 & -1 & 1 \\ -1 & -1 & 1 & 1 \\ 1 & -1 & -1 & 1 \end{pmatrix}\begin{Bmatrix} (1) \\ a \\ b \\ ab \end{Bmatrix} \\ &= \frac{1}{n2^2}\begin{Bmatrix} (1) + a + \text{ab} + b \\ -(1) + a + \text{ab} - b \\ -(1) - a + \text{ab} + b \\ (1) - a + \text{ab} - b \end{Bmatrix} = \frac{1}{n2^2}\begin{Bmatrix} C_I \\ C_A \\ C_B \\ C_{AB} \end{Bmatrix} = \frac{1}{2}\begin{Bmatrix} \text{Effect}_I \\ \text{Effect}_A \\ \text{Effect}_B \\ \text{Effect}_{AB} \end{Bmatrix} \end{aligned} \tag{4.73}$$

Using Eqs. (4.71), (4.73), a4.and (4.63), the predicted value of the response is obtained from

$$\widehat{y} = \overline{y} + \frac{1}{2}(\text{Effect}_A x_1 + \text{Effect}_B x_2 + \text{Effect}_{AB} x_1 x_2) \tag{4.74}$$

and x_1 and x_2 are the coded values.

The residuals ε are obtained from

$$\varepsilon = y - \widehat{y}$$

where y is the corresponding measured response value at each combination of the coded values of x_j for each replicate.

We now consider the 2^3-factorial experiment. In this case, signs given in Table 4.12 become those shown in Table 4.17 and the analysis of variance becomes that shown in Table 4.18. Then, Eq. (4.69) becomes

Table 4.17 Signs of the terms appearing in the Contrasts C_γ for a 2^3-factorial experiment

Treatment combination	γ						
	A	B	AB	C	AC	BC	ABC
(1)	−	−	+	−	+	+	−
a	+	−	−	−	−	+	+
b	−	+	−	−	+	−	+
ab	+	+	+	−	−	−	−
c	−	−	+	+	−	−	+
ac	+	−	−	+	+	−	−
bc	−	+	−	+	−	+	−
abc	+	+	+	+	+	+	+

Table 4.18 2^3-factorial analysis of variance, where $c_0 = 2^3(n-1)$ and $d_0 = 2^3 n - 1$

Source of variation	Degrees of freedom	Sum of squares	Mean square	F value	p-value[a]
Factor A	1	SS_A	$\mathrm{MS}_A = SS_A$	$F_A = \dfrac{\mathrm{MS}_A}{\mathrm{MS}_E}$	$\Psi_f(F_A, 1, c_0)$
Factor B	1	SS_B	$\mathrm{MS}_B = SS_B$	$F_B = \dfrac{\mathrm{MS}_B}{\mathrm{MS}_E}$	$\Psi_f(F_B, 1, c_0)$
Factor C	1	SS_C	$\mathrm{MS}_C = SS_C$	$F_C = \dfrac{\mathrm{MS}_C}{\mathrm{MS}_E}$	$\Psi_f(F_C, 1, c_0)$
Interaction AB	1	SS_{AB}	$\mathrm{MS}_{AB} = SS_{AB}$	$F_{AB} = \dfrac{\mathrm{MS}_{AB}}{\mathrm{MS}_E}$	$\Psi_f(F_{AB}, 1, c_0)$
Interaction AC	1	SS_{AC}	$\mathrm{MS}_{AC} = SS_{AC}$	$F_{AC} = \dfrac{\mathrm{MS}_{AC}}{\mathrm{MS}_E}$	$\Psi_f(F_{AC}, 1, c_0)$
Interaction BC	1	SS_{BC}	$\mathrm{MS}_{BC} = SS_{BC}$	$F_{BC} = \dfrac{\mathrm{MS}_{BC}}{\mathrm{MS}_E}$	$\Psi_f(F_{BC}, 1, c_0)$
Interaction ABC	1	SS_{ABC}	$\mathrm{MS}_{ABC} = SS_{ABC}$	$F_{ABC} = \dfrac{\mathrm{MS}_{ABC}}{\mathrm{MS}_E}$	$\Psi_f(F_{ABC}, 1, c_0)$
Error	c_0	SS_E	$\mathrm{MS}_E = \dfrac{SS_E}{c_0}$		
Total	d_0	SS_T			

[a] Ψ_f is determined from Eq. (2.120)

$$\boldsymbol{C} = \begin{Bmatrix} \mathrm{C}_A \\ \mathrm{C}_B \\ \mathrm{C}_{AB} \\ \mathrm{C}_C \\ \mathrm{C}_{AC} \\ \mathrm{C}_{BC} \\ \mathrm{C}_{ABC} \end{Bmatrix} \quad \boldsymbol{g} = \begin{Bmatrix} (1) \\ a \\ b \\ ab \\ c \\ ac \\ bc \\ abc \end{Bmatrix}$$

$$\boldsymbol{h} = \begin{pmatrix} -1 & -1 & 1 & -1 & 1 & 1 & -1 \\ 1 & -1 & -1 & -1 & -1 & 1 & 1 \\ -1 & 1 & -1 & -1 & 1 & -1 & 1 \\ 1 & 1 & 1 & -1 & -1 & -1 & -1 \\ -1 & -1 & 1 & 1 & -1 & -1 & 1 \\ 1 & -1 & -1 & 1 & 1 & -1 & -1 \\ -1 & 1 & -1 & 1 & -1 & 1 & -1 \\ 1 & 1 & 1 & 1 & 1 & 1 & 1 \end{pmatrix} \tag{4.75}$$

Upon performing the matrix multiplication given by Eq. (4.70), we obtain

$$\begin{Bmatrix} \mathrm{C}_A \\ \mathrm{C}_B \\ \mathrm{C}_{AB} \\ \mathrm{C}_C \\ \mathrm{C}_{AC} \\ \mathrm{C}_{BC} \\ \mathrm{C}_{ABC} \end{Bmatrix} = \begin{Bmatrix} -(1) + a + ab + abc + ac - b - bc - c \\ -(1) - a + ab + abc - ac + b + bc - c \\ (1) - a + ab + abc - ac - b - bc + c \\ -(1) - a - ab + abc + ac - b + bc + c \\ (1) - a - ab + abc + ac + b - bc - c \\ (1) + a - ab + abc - ac - b + bc - c \\ -(1) + a - ab + abc - ac + b - bc + c \end{Bmatrix} \tag{4.76}$$

Then the sum of squares and the effects are obtained from

$$\begin{aligned} \mathrm{Effect}_\gamma &= \frac{1}{2^2 n}\mathrm{C}_\gamma \\ SS_\gamma &= \frac{1}{2^3 n}C_\gamma^2 \qquad \gamma = A, B, C, AB, AC, BC, ABC \end{aligned} \tag{4.77}$$

By modifying Eq. (4.67) to include the third factor C, we obtain

$$SS_T = \sum_{i=1}^{2}\sum_{j=1}^{2}\sum_{k=1}^{2}\sum_{m=1}^{n} y_{ijkm}^2 - 2^3 n\bar{y}^2 \tag{4.78}$$

where $\bar{y}$ is now given by

$$\bar{y} = \frac{1}{2}\text{Effect}_I \tag{4.79}$$

and Effect_I is given by

$$\begin{aligned}\text{Effect}_I &= \frac{1}{n2^2}C_I \\ C_I &= (1) + a + b + ab + c + ac + bc + abc\end{aligned} \tag{4.80}$$

In Eq. (4.78), the subscript $i = 1$ denotes that factor A is at its low level and the subscript $i = 2$ denotes the factor A is at its high level. Similarly, for the subscript j and factor B and for the subscript k and the factor C. The quantity SS_T has $2^3n - 1$ degrees of freedom. The error sum of squares becomes

$$SS_E = SS_T - SS_A - SS_B - SS_{AB} - SS_C - SS_{AC} - SS_{BC} - SS_{ABC} \tag{4.81}$$

Following the procedure to obtain Eq. (4.74), the predicted value of the response is obtained from

$$\begin{aligned}\hat{y} = \bar{y} + \frac{1}{2}(&\text{Effect}_A x_1 + \text{Effect}_B x_2 + \text{Effect}_{AB} x_1 x_2 + \text{Effect}_C x_3 \\ &+\text{Effect}_{AC} x_1 x_3 + \text{Effect}_{BC} x_2 x_3 + \text{Effect}_{ABC} x_1 x_2 x_3)\end{aligned} \tag{4.82}$$

where x_1, x_2, and x_3 are the coded values. In practice, the evaluation for the predicted response given by Eq. (4.82) is modified to only include those effects that typically have a p-value less than 0.01.

The residuals ε are obtained from

$$\varepsilon = y - \hat{y} \tag{4.83}$$

where y is the corresponding measured response value at each combination of the coded values of x_j for each replicate.

These results are illustrated with the following example, which uses interactive graphic IG4-4.

Example 4.4

Consider the data from a 2^3-factorial experiment with two replicates ($n = 2$), which appear in the fourth and fifth columns of Table 4.19. This table also includes several additional quantities computed from these data. These additional quantities are used to create Table 4.20.

Table 4.19 Data for a 2^3-factorial experiment with some additional quantities derived from these data as indicated in Eqs. (4.76) and (4.77) and where it is found that $\bar{y} = 777.56$

$A \rightarrow x_1$	$B \rightarrow x_2$	$C \rightarrow x_3$	y_{ijk1}	y_{ijk2}	$y_{ijk\bullet}$	Contrast	γ = Effect$_\gamma$
−1	−1	−1	545	622	$(1) = 1167$		
1	−1	−1	687	628	$a = 1315$	$C_A = -609$	$A = -76.13$
−1	1	−1	612	552	$b = 1164$	$C_B = -73$	$B = -9.13$
1	1	−1	642	681	$ab = 1323$	$C_{AB} = -75$	$AB = -9.38$
−1	−1	1	1059	1036	$c = 2095$	$C_C = 2503$	$C = 312.9$
1	−1	1	794	886	$ac = 1680$	$C_{AC} = -1223$	$AC = -152.9$
−1	1	1	1026	1073	$bc = 2099$	$C_{BC} = -83$	$BC = -10.4$
1	1	1	761	837	$abc = 1589$	$C_{ABC} = -97$	$ABC = -12.13$

Table 4.20 Analysis of variance for the data in Table 4.19

Source	Degrees of freedom	Sum of squares	Mean square	*F* value	*p*-value
A	1	23180.1	23180.1	11.77	0.008944
B	1	333.063	333.063	0.1691	0.6917
AB	1	351.563	351.563	0.1785	0.6838
C	1	391,563	391,563	198.8	6.217×10^{-7}
AC	1	93483.1	93483.1	47.47	1.258×10^{-4}
BC	1	430.563	430.563	0.2186	0.6526
ABC	1	588.063	588.063	0.2986	0.5997
Error	8	15754.5	1969.31		
Total	15	525,684.			

From the *p*-values in Table 4.20, we see that factors *A* and *C* and the interaction of these two factors are dominant. Therefore, our model for the predicted response given by Eq. (4.82) can be reduced to

$$\hat{y} = \bar{y} + \frac{1}{2}\left(\text{Effect}_A x_1 + \text{Effect}_C x_3 + \text{Effect}_{AC} x_1 x_3\right) \quad \text{(a)}$$

Using Eq. (a), we obtain the residuals listed in Table 4.21.

Table 4.21 Residuals for the data in Table 4.19 using the predicted model given by Eq. (a)

Factor levels	Residuals	
	m = 1	m = 2
(1)	−37.8	39.3
a	27.5	−31.5
b	29.3	−30.8
ab	−17.5	21.5
c	10.5	−12.5
ac	−25.5	66.5
bc	−22.5	24.5
abc	−58.5	17.5

4.6 Exercises

Section 4.2

4.1. Perform an analysis of variance on the data shown in Table 4.22, which were obtained from a single factor experiment at five treatment levels. From the analysis:

(a) Determine if at least one treatment mean is different from the other treatment means.
(b) Determine if the residuals have a normal distribution.
(c) Determine the confidence interval for the mean value at each treatment level for $\alpha = 0.05$.
(d) For $\alpha = 0.05$, determine if the means for treatment levels 2 and 3 are different and if treatment levels 4 and 5 are different.

Table 4.22 Data for Exercise 4.1

A_1	A_2	A_3	A_4	A_5
544	586	636	410	566
462	590	621	452	623
450	511	507	525	528
739	583	573	444	613
497	639	646	420	661
642	527	668	549	682

Table 4.23 Data for Exercise 4.2

	B_1	B_2	B_3	B_4
A_1	871	988	1039	1080
A_2	911	1004	1099	1165
A_3	923	999	1085	1152
A_4	1009	1094	1164	1222
A_5	1028	1129	1193	1242

Table 4.24 Data for Exercise 4.3

	B_1	B_2	B_3
A_1	5.232, 5.331, 7.421	4.814, 7.74, 6.497	7.487, 4.869, 5.617
A_2	4.484, 4.352, 4.825	5.023, 5.551, 6.178	4.055, 5.133, 4.352
A_3	7.399, 5.606, 4.418	6.2, 6.365, 8.125	6.068, 5.639, 4.539
A_4	9.368, 8.125, 8.642	7.773, 7.157, 10.05	9.841, 8.51, 9.06

Section 4.3

4.2. Five different models of air conditioners are compared using their power consumption as the metric. It has been decided to use as a blocking variable the percentage humidity by determining their power consumption at four different humidity levels. The results of the test are shown in Table 4.23.

(a) Determine if there is at least one air conditioner that is different from the others.
(b) Is humidity an influential blocking factor?

Section 4.3

4.3. The strength of a solder joint is being examined by considering four different solder compositions and three different vendors to apply the solder. At each combination of vendor and solder composition, the strength of three samples will be tested. The result of this experiment is shown in Table 4.24. Use interactive graphic IG4-3 to analyze the data and then discuss the results.

Appendix A: Moment Generating Function

A.1 Moment Generating Function

The moment generating function $M_x(t)$ is defined as $E(e^{tx})$; thus, from Eq. (2.7)

$$M_x(t) = E(e^{tx}) = \int_{-\infty}^{\infty} e^{tx} f(x) dx \tag{A.1}$$

We note that

$$\begin{aligned} \frac{dM_x}{dt} &= M_x'(t) = \int_{-\infty}^{\infty} x e^{tx} f(x) dx \\ M_x'(0) &= \int_{-\infty}^{\infty} x f(x) dx = E(X) \\ \frac{d^2 M_x}{dt^2} &= M_x''(t) = \int_{-\infty}^{\infty} x^2 e^{tx} f(x) dx \\ M_x''(0) &= \int_{-\infty}^{\infty} x^2 f(x) dx = \mathrm{Var}(X) + [E(X)]^2 \end{aligned} \tag{A.2}$$

where we have used Eqs. (2.7) and (2.8).

It is noted that if

$$U = \sum_{i=1}^{n} X_i \tag{A.3}$$

E. B. Magrab, *Engineering Statistics*, https://doi.org/10.1007/978-3-031-05010-7

then

$$\begin{aligned} M_U(t) &= E\left(e^{tU}\right) = E\left(\exp\left(t\sum_{i=1}^{n} X_i\right)\right) = E\left(e^{tX_1}\right)E\left(e^{tX_2}\right)\cdots E\left(e^{tX_n}\right) \\ &= M_{X_1}(t)M_{X_2}(t)\cdots M_{X_n}(t) \end{aligned} \tag{A.4}$$

A.2 Moment Generating Function of the Chi Square Distribution

To find the moment generating function of the chi square distribution, we use the probability density function given by Eq. (2.52); that is

$$\begin{aligned} f_\chi(x) &= \frac{2^{-\nu/2}}{\Gamma(\nu/2)} x^{\nu/2-1} e^{-x/2} \quad x > 0 \\ &= 0 \quad x \le 0 \end{aligned} \tag{A.5}$$

Substituting Eqs. (A.5) into Eq. (A.1), the moment generating function for the chi square distribution is

$$\begin{aligned} M_x(t) &= \frac{2^{-\nu/2}}{\Gamma(\nu/2)} \int_0^\infty x^{\nu/2-1} e^{x(2t-1)/2} dx \\ &= (1-2t)^{-\nu/2} \quad t < 1/2 \quad \nu > 0 \end{aligned} \tag{A.6}$$

Let X_i, $i = 1, 2, \ldots, n$, be normally distributed independent random variables; that is, $X_i \sim N(0,1)$. We shall show by using the moment generating function that the following sum is a chi square variable

$$V = \sum_{i=1}^{n} X_i^2 \tag{A.7}$$

We start by considering the moment generating function for one term, say $X_i^2 = X^2$. Since $X_i \sim N(0,1)$ the probability density function is given by Eq. (2.14); that is

$$f_N(x) = \frac{1}{\sqrt{2\pi}} e^{-x^2/2} \tag{A.8}$$

Then, substituting Eq. (A.8) into Eq. (A.1), the moment generating function becomes

$$M_x(t) = \frac{1}{\sqrt{2\pi}} \int_{-\infty}^{\infty} e^{tx^2} e^{-x^2/2} dx = \frac{1}{\sqrt{2\pi}} \int_{-\infty}^{\infty} e^{(2t-1)x^2/2} dx = (1-2t)^{-1/2} \tag{A.9}$$

Using Eq. (A.4), we find with X_i replaced by X_i^2 that

$$\begin{aligned} M_V(t) &= E\left(e^{tV}\right) = E\left(e^{tX_1^2}\right)E\left(e^{tX_2^2}\right)\cdots E\left(e^{tX_n^2}\right) \\ &= M_{X_1^2}(t)M_{X_2^2}(t)\cdots M_{X_n^2}(t) \\ &= (1-2t)^{-1/2}(1-2t)^{-1/2}\cdots(1-2t)^{-1/2} \\ &= (1-2t)^{-n/2} \end{aligned} \tag{A.10}$$

However, this is the same moment generating function as for the chi square distribution given by Eq. (A.6) with $\nu = n$. Thus, $V \sim \chi_n^2$; that is, V is a chi square distribution.

We can use of Eq. (A.10) to make the following statements.

(a) If X and Y are independent random variables and $X \sim \chi_m^2$ and $Y \sim \chi_n^2$ then $X + Y \sim \chi_{m+n}^2$ since, from Eqs. (A.4) and (A.10),

$$M_{x+y}(t) = M_x(t)M_y(t) = (1-2t)^{-m/2}(1-2t)^{-n/2} = (1-2t)^{-(m+n)/2} \tag{A.11}$$

(b) If $Z = X + Y$ where $X \sim \chi_m^2$ and $Z \sim \chi_n^2$ with $n > m$, then $Y \sim \chi_{n-m}^2$ since

$$\begin{aligned} M_{x+y}(t) &= M_x(t)M_y(t) \\ (1-2t)^{-n/2} &= (1-2t)^{-m/2}M_y(t) \\ M_y(t) &= (1-2t)^{-(n-m)/2} \end{aligned} \tag{A.12}$$

which is a chi square distribution.

A.3 Independence of Mean and Variance for Normally Distributed Independent Random Variables

For a sample whose random variables are independent and for which $X_i \sim N(0,1)$, we shall show that the sample mean and sample variance are independent. The mean and variance, respectively, are

$$
\begin{aligned}
\overline{x} &= \frac{1}{n}\sum_{i=1}^{n} x_i \\
s^2 &= \frac{1}{n-1}\sum_{i=1}^{n} (x_i - \overline{x})^2
\end{aligned}
\tag{A.13}
$$

To show independence, it is sufficient to show that X_i and $X_i - \overline{X}$ are independent. First, we obtain the moment generating function of $\overline{X}$. Thus, from Eqs. (A.1) and (A.8), we find that

$$M_{x_i/n}(t) = \frac{1}{\sqrt{2\pi}} \int_{-\infty}^{\infty} e^{tx/n - x^2/2} dx = e^{t^2/(2n^2)} \tag{A.14}$$

Using Eq. (A.4) with $U = \overline{X}$, we find that

$$M_{\overline{x}}(t) = e^{t^2/(2n^2)} e^{t^2/(2n^2)} \cdots e^{t^2/(2n^2)} = e^{t^2/(2n)} \tag{A.15}$$

and from Eq. (A.2),

$$
\begin{aligned}
M'_{x_i/n}(t) &= \frac{te^{t^2/(2n)}}{n} \to M'_x(0) = 0 \\
M''_{x_i/n}(t) &= \frac{e^{t^2/(2n)}}{n} + \frac{t^2 e^{t^2/(2n)}}{n^2} \to M''_x(0) = \frac{1}{n}
\end{aligned}
\tag{A.16}
$$

Hence, $\overline{X} \sim N(0, 1/n)$.

We note that $\overline{X} - X_j$ can be written as

$$\overline{X} - X_j = \frac{1}{n}\{X_1 + X_2 + \cdots + X_{j-1} + X_{j+1} \cdots X_n\} - \frac{n-1}{n} X_j \tag{A.17}$$

The moment generating function for each term inside the braces is given by Eq. (A.14). The moment generating function for the right-most term in Eq. (A.17) is

$$M_{x_j(n-1)/n}(t) = \frac{1}{\sqrt{2\pi}} \int_{-\infty}^{\infty} e^{-tx(n-1)/n - x^2/2} dx = e^{t^2(n-1)^2/(2n^2)} \tag{A.18}$$

Then the moment generating function for $\overline{X} - X_i$ is

$$M_{\overline{X}-X_j}(t) = \left(e^{t^2/(2n^2)}\right)^{n-1} e^{t^2(n-1)^2/(2n^2)} = e^{t^2(n-1)/(2n)} \tag{A.19}$$

Therefore,

$$\begin{aligned} M'_{\overline{X}-X_j}(t) &= \frac{n-1}{n} t e^{t^2(n-1)/(2n)} \rightarrow M'_{\overline{X}-X_j}(0) = 0 \\ M''_{\overline{X}-X_j}(t) &= \left(\frac{n-1}{n} + \frac{(n-1)^2}{n^2} t^2\right) e^{t^2(n-1)/(2n)} \rightarrow M''_{\overline{X}-X_j}(0) = \frac{n-1}{n} \end{aligned} \tag{A.20}$$

Hence, $\overline{X}-X_j \sim N(0,(n-1)/n)$; that is, $E(\overline{X}-X_j)=0$ and $\mathrm{Var}(\overline{X}-X_j)=(n-1)/n$.

We shall show the independence of X_i and $X_i - \overline{X}$ by examining their covariance. We note that (recall Eq. (1.21) of which this is an extension)

$$\begin{aligned} \mathrm{Cov}(\overline{X}, X_j) &= \mathrm{Cov}\left(\frac{1}{n}\sum_{i=1}^{n} X_i, X_j\right) = \frac{1}{n}\sum_{i=1}^{n} \mathrm{Cov}(X_i, X_j) \\ &= \frac{1}{n}\mathrm{Cov}(X_j, X_j) = \frac{1}{n}\mathrm{Var}(X_j) = \frac{1}{n} \end{aligned} \tag{A.21}$$

since it was assumed that X_i are independent random variables; that is, $\mathrm{Cov}(X_i, X_j) = 0$, $i \neq j$, and $X_i \sim N(0,1)$; that is, $\mathrm{Var}(X_j) = 1$. Then,

$$\begin{aligned} \mathrm{Cov}(\overline{X}, X_j - \overline{X}) &= \mathrm{Cov}(\overline{X}, X_j) - \mathrm{Cov}(\overline{X}, \overline{X}) \\ &= \mathrm{Cov}(\overline{X}, X_j) - \mathrm{Var}(\overline{X}) \\ &= \frac{1}{n} - \frac{1}{n} = 0 \end{aligned} \tag{A.22}$$

where we have used Eq. (A.21) and from Eq. (A.16) used the fact that $\mathrm{Var}(\overline{X}) = 1/n$. Since the covariance is zero these quantities are independent. This implies that $\overline{X}$ and $(\overline{X} - X_j)^2$ are also independent and therefore $\overline{X}$ and s^2 are independent.

Bibliography

DeCoursey WJ (2003) Statistics and probability for engineering applications. Elsevier, Woburn, MA
Devore JL (2012) Probability and statistics for engineering and the sciences, 8th edn. Brooks/Cole, Boston, MA
Hayter A (2012) Probability and statistics for engineers and scientists, 4th edn. Brooks/Cole, Boston, MA
Montgomery DC (2013) Design and analysis of experiments, 8th edn. Wiley, Hoboken, NJ
Montgomery DC, Runger GC (2018) Applied statistics and probability for engineers, 7th edn. Wiley, Hoboken, NJ
Navidi W (2011) Statistics for engineers and scientists, 3rd edn. McGraw-Hill, New York, NY
Ross SM (2014) Introduction to probability and statistics for engineers and scientists, 5th edn. Academic Press, New York, NY
Soong TT (2004) Fundamentals of probability and statistics for engineers. Wiley, Chichester
Walpole RE, Myers RH, Myers SL, Ye K (2012) Probability & statistics for engineers & scientists, 9th edn. Prentice Hall, New York, NY

Internet

Engineering Statistics Handbook, produced by the National Institute of Standard and Technology (NIST): https://www.itl.nist.gov/div898/handbook/

E. B. Magrab, *Engineering Statistics*, https://doi.org/10.1007/978-3-031-05010-7

Index

E. B. Magrab, *Engineering Statistics*, https://doi.org/10.1007/978-3-031-05010-7

Printed in the United States
by Baker & Taylor Publisher Services